Design of Hydrodynamic Machines

Design of Hydrodynamic Machines

Pumps and Hydro-Turbines

Getu Hailu
Michal Varchola
Peter Hlbocan

CRC Press
Taylor & Francis Group
Boca Raton London New York

CRC Press is an imprint of the
Taylor & Francis Group, an **informa** business

First edition published 2022
by CRC Press
6000 Broken Sound Parkway NW, Suite 300, Boca Raton, FL 33487-2742

and by CRC Press
4 Park Square, Milton Park, Abingdon, Oxon, OX14 4RN

CRC Press is an imprint of Taylor & Francis Group, LLC

Library of Congress Cataloging-in-Publication Data
Names: Hailu, Getu, author. | Varchola, Michal, author. | Hlbocan, Peter, author.
Title: Design of hydrodynamic machines : pumps and hydro-turbines /
Getu Hailu, Michal Varchola, Peter Hlbocan.
Description: First edition. | Boca Raton, FL: CRC Press, 2022. |
Includes bibliographical references and index. |
Summary: "Design of Hydrodynamic Machines provides a broad, yet concise, theoretical background on the relationship between fluid dynamics and geometry. It covers the most important types of turbomachinery used in power generation industrial processes, utilities, and the oil and gas industry. Intended for final year undergraduates and postgraduates in mechanical, civil, and aeronautical engineering, the book will also be useful for those involved in the hydraulic design, analysis, and testing of turbomachinery. The book offers guidance on the hydraulic design aspect of different parts of turbomachinery, such as impellers, diffusers, volute casing, inlet and outlets"– Provided by publisher.
Identifiers: LCCN 2021049580 (print) | LCCN 2021049581 (ebook) |
ISBN 9780367439613 (hbk) | ISBN 9781032229294 (pbk) |
ISBN 9781003007142 (ebk)
Subjects: LCSH: Turbomachines.
Classification: LCC TJ266 .H34 2022 (print) |
LCC TJ266 (ebook) | DDC 621.406–dc23/eng/20220103
LC record available at https://lccn.loc.gov/2021049580
LC ebook record available at https://lccn.loc.gov/2021049581

ISBN: 978-0-367-43961-3 (hbk)
ISBN: 978-1-032-22929-4 (pbk)
ISBN: 978-1-003-00714-2 (ebk)

DOI: 10.1201/9781003007142

Typeset in Times
by Newgen Publishing UK

Contents

Nomenclature

Symbol	Unit	Description
A	[m^2]	Flow area
b	[m]	Meridional section width, span
c	[ms^{-1}]	Absolute speed, speed of sound, velocity
c_m	[ms^{-1}]	Meridional velocity
c_u	[ms^{-1}]	Tangential component of absolute velocity
c_r	[ms^{-1}]	Radial component of absolute velocity
C_D	[-]	Drag coefficient
C_L	[-]	Lift coefficient
d_n	[m]	Impeller hub diameter
D, D_A	[m]	Diameter
e	[mm]	Wall thickness of a penstock
e_z	[Jkg^{-1}]	Losses
F	[N]	Force
f	[-]	Friction factor
F_A	[N]	Axial thrust
G	[Pa]	Modulus of elastisity in shear
g	[ms^{-2}]	Acceleration due to gravity
H	[m]	Head
H_S	[m]	Suction height
h	[m]	Height
K	[Pa]	Bulk modulus
L	[m]	Blade span, characteristic length
l	[m]	Length of chord
L^*	[-]	Location of maximum camber
$\frac{L^*}{l}$	[-]	Ratio of distance of maximum camber to chord length
m	[m]	Maximum camber
$\frac{m}{l}$	[-]	Ratio of maximum camber to chord length
M_k	[Nm]	Shaft torque
n, N	[s^{-1}]	Speed
n_b	[-]	Specific speed
n_s	[min^{-1}]	Dimensional specific speed
$NPSH$	[m]	Net positive suction head
p	[Pa]	Pressure
p_A	[Pa]	Absolute pressure
p_w	[Pa]	Vapor pressure
P	[W]	Power
Q	[m^3/s]	Volumetric flow rate

Symbol	Unit	Description
$\dot{m}$	$[kgs^{-1}]$	Mass flow rate
R, r	[m]	Radius
R_{os}	[m]	Place pitch radius
R_r	[m]	Radius of a spiral
s	[m]	Blade thickness
T		Temperature
t	[-]	Pitch
t^*	[m]	Thickness at the maximum camber location
$\frac{t^*}{l}$	[-]	Ratio of thickness at maximum camber to chord length
u	$[ms^{-1}]$	Tangential velocity
v	$[ms^{-1}]$	Average velocity
w	$[ms^{-1}]$	Relative velocity
w_u	$[ms^{-1}]$	Tangential component of the relative velocity
w_m	$[ms^{-1}]$	Meridional component of the relative velocity
w_∞	$[ms^{-1}]$	Undisturbed mean relative velocity
Y	$[Jkg^{-1}]$	Specific energy
ΔY	$[Jkg^{-1}]$	Change in specific energy
z	[m]	Elevation, number of blades
z_{sg}	[m]	Elevation on the suction side of a pump
z_{vg}	[m]	Elevation on the discharge side of a pump

Greek Symbols

Symbol	Unit	Description
$\propto$	[rad]	Angle between the absolute velocity and the tangential velocity
β	[rad]	Angle between the relative velocity and the tangential velocity
β	[1/°C]	Coefficient of volume expansion
β_∞	[rad]	Angle between undisturbed mean relative velocity and tangential velocity
δ	[rad]	Angle of attack
Δ	[-]	Change
φ	[-]	Contraction coefficient
γ	[$kgm^{-2}s^{-2}$]	Specific weight
Γ	[m^2s^{-1}]	Circulation
Γ_z	[m^2s^{-1}]	Circulation around a single blade
η	[-], [%]	Efficiency
ν	[m^2s^{-1}]	Kinematic viscosity
π	[-]	Pi
ρ	[kg m^{-3}]	Density
σ	[Pa]	Stress
σ_p	[-]	Thoma's cavitation coefficient
σ_Y	[-]	Correction factor
μ	[Pa.s]	Dynamic viscosity
τ	[Pa]	Shear stress
θ	[rad]	Angle
ω	[$rads^{-1}$]	Angular velocity
ξ	[-]	Minor loss coefficient
ζ	[$rads^{-1}$]	Vorticity vector

Authors' Biographies

Getu Hailu is currently Associate Professor of Mechanical Engineering at the University of Alaska Anchorage. He has more than 20 years of experience in research and teaching. He designed, developed, and has been teaching turbomachinery courses at the University of Alaska Anchorage at both the graduate and undergraduate levels. He has supervised graduate and undergraduate research students, mainly in the thermo-fluids area. He is author/coauthor of more than 40 refereed publication. He is a member of ASME, ASHRAE, and IBPSA.

Peter Hlbocan is currently a research engineer – Simulation Specialist with ZTS Research and Development (ZTS VaV, a.s.). His main duties include CFD simulations in the area of hydro energy. He has more than ten years of experience in CFD flow modeling. He has worked on several industry-sponsored research projects focusing mainly on hydraulic performance improvements (modifications) and other hydraulic parameters. His work also includes CFD modeling of water pumps (radial, mixed-flow) and hydraulic turbines. He has also experience in teaching thermo-fluids courses and supervision of theses at Slovak University of Technology in Bratislava. He is author/coauthor of more than 30 refereed publications.

Michal Varchola has been teaching and conducting research on turbomachinery for more than 40 years. He has published 5 monographs, 4 textbooks, 4 books, and more than 150 articles in journals and refereed conference proceedings. He has supervised 12 theses. He has been a lead investigator for numerous industry-sponsored research. Varchola is a former chair of the Hydraulic and Pneumatic Equipment Machinery Department, Dean of Faculty of Mechanical Engineering, and Vice Rector of Slovak Technical University.

Preface

Design of Hydrodynamic Machines: Pumps and Hydro-Turbines is intended as a textbook for both upper-level undergraduate and graduate students. Primarily it is intended for students enrolled in mechanical, civil, and aerospace engineering programs. However, students in other programs such as chemical engineering and manufacturing and design engineering will also find it useful.

Most turbomachinery books cover a wide range of machines, such as pumps, fans, compressors, turbines (steam/gas and hydro), and wind turbines. Because of the wide range of coverage, there is limited space to go into the topics in detail. This textbook tries to address this shortcoming. It focuses on hydrodynamic machines, mainly pumps and hydro-turbines. The design and use of conventional types of hydrodynamic machines used in power generation, industrial processes, utilities, and oil and gas industry are covered in detail. The book gives detailed, easy-to-follow guidance on how to perform hydraulic design of important components of these machines; how to establish performance characteristics; and the procedures for testing and evaluation of these machines. The book also describes cavitation and its effects on the performance of a hydrodynamic machine. It presents methods of designing the suction side of the piping system to prevent cavitation.

One chapter is dedicated to the application of computational fluid dynamics (CFD) to the design and analysis of turbomachines. Three problems are solved illustrating CFD application to the design of piping components, an impeller with a diffuser, and a turbine draft tube.

Example problems are solved to illustrate design problems. Additional exercise problems are provided at the end of the chapters.

1 Introduction

1.1 FUNDAMENTAL PRINCIPLES

Turbomachines are among the most widely used devices in the world. Pumps, compressors, turbines, and fans are all collectively known as turbomachines. In turbomachinery, energy transfer occurs between a continuously flowing fluid (liquid, steam, gas) and a rotor by the dynamic action of moving blade rows. The word *turbo* or *turbinis* is a Latin word meaning "that which spins or whirls around." Turbomachines can generally be classified into two categories. Turbomachines that transfer energy from a continuously flowing fluid to a blade rotor are known as turbines. Those devices transferring energy from a rotor to a continuously flowing fluid are known as compressors (air), pumps (for liquids), or fans (air). Compressors result in a high-pressure rise compared to fans. These devices use shaft power to increase fluid pressure or head. Some authors define *open turbomachines* that include wind turbines, propellers, and fans without shroud. Whereas pumps, compressors, and fans increase the pressure of the fluid, turbines produce power by rotating the rotor of a generator.

In designing turbomachines, fundamental laws of conservation of mass, momentum, and energy are applied. Several turbomachinery design methodologies have been developed, ranging from simple one-dimensional (1D) analysis to 3D flow analysis. 1D methods that introduce simplifying hypotheses, 2D methods that are based on potential flow theory, and more rigorous 3D design methodologies utilizing numerical methods of solving governing fluid flow equations have all been used to design turbomachines. Modern computational fluid dynamics (CFD) software gives detailed information on velocity and pressure distributions in turbomachines with sufficient accuracy.

1.2 TURBOMACHINERY CLASSIFICATION

Turbomachines can be classified based on the type of working fluid they handle (Figure 1.1). Turbomachines that handle incompressible working fluid are known as hydraulic turbomachines while those handling compressible working fluid are known as thermal turbomachines. In general, if the working fluid is liquid, it can be considered incompressible with constant density. If the working fluid is gas it can be treated as incompressible if the Mach number is less than 0.3. At higher Mach numbers, however, the gas should be considered as compressible. An example of a thermal turbomachine is a gas turbine (Figure 1.2), which contains a compressor, combustion chamber, and turbine. The compressor pulls air from the surroundings and compresses it in several stages. In the combustion chamber, heat is supplied to

DOI: 10.1201/9781003007142-1

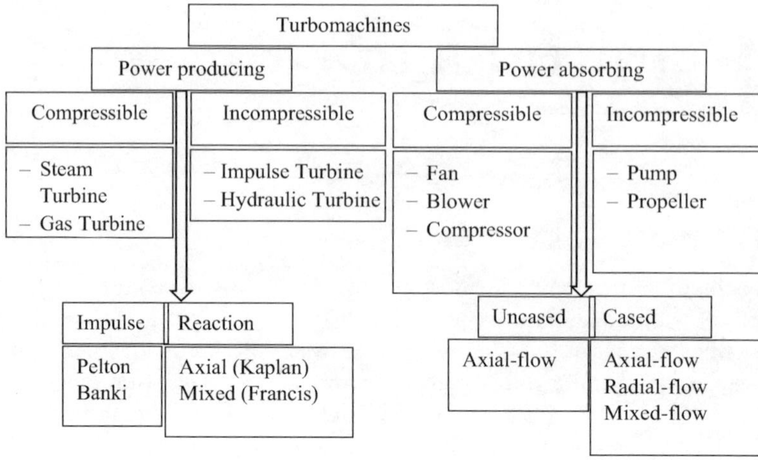

FIGURE 1.1 Classification of turbomachines.

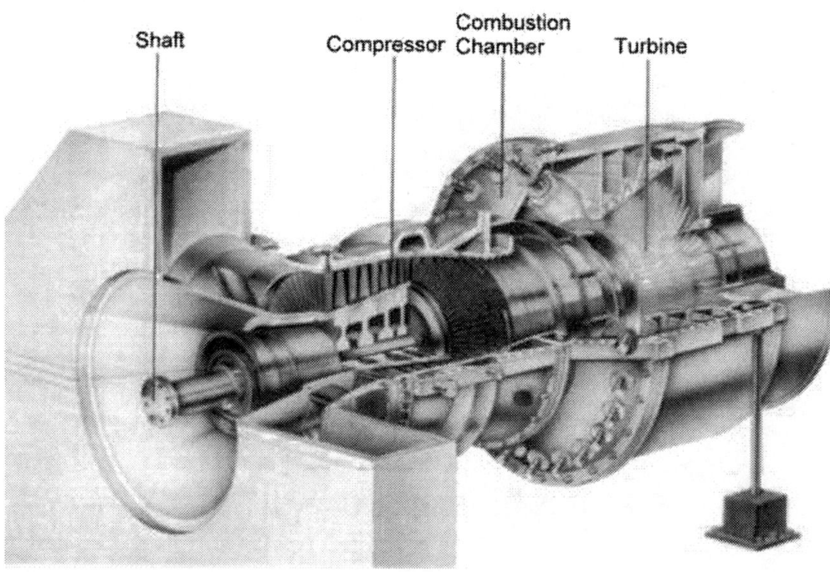

FIGURE 1.2 A gas turbine (source: www.wartsila.com/).

the flow by almost isobaric combustion process. The heated air in the combustion chamber then passes through the turbine, where energy is removed from the fluid and is carried off in the form of mechanical work by the drive shaft.

Turbomachines can also be classified according to the direction of fluid flow (the path the fluid takes) through the passages of a rotor as axial flow, radial flow, and mixed-flow types.

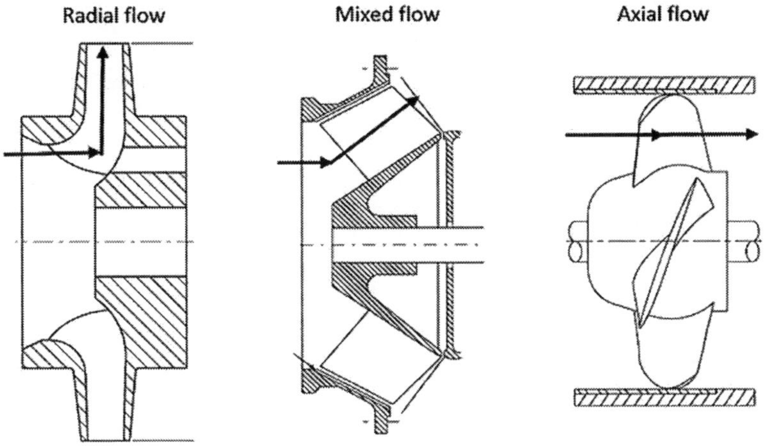

FIGURE 1.3 Cross-sectional view of radial, mixed-flow, and axial pumps (www.pumpsandsystems.com/).

In *radial flow turbomachines* the flow direction is predominantly in a plane perpendicular to the axis of rotation. Figure 1.3 (left) shows a cross-sectional view of a radial pump. Note that fluid intake is in the direction parallel to the axis of rotation and the rotor blades push the fluid radially (i.e., in the plane perpendicular to the axis of rotation).

In *mixed-flow turbomachines* (Figure 1.3 center) the predominant flow direction at the impeller outlet results in radial and axial velocity components.

In *axial flow turbomachines* (Figure 1.3 right) the flow direction is predominantly parallel to the axis of rotation.

1.3 DIMENSIONS AND UNITS USED IN TURBOMACHINERY

There are two systems of units widely used: the International System of Units (SI) and Imperial Units (IP). The SI unit is also commonly known as the metric unit. The SI unit was adopted in 1960 at the General Conference on Weights and Measures. The SI system of units is based on the meter-kilogram-second (MKS), also known as MLT (mass, length, time) system, in terms of the most-used primary dimensions. The SI unit is simple and logical; it is based on decimal relationships between units. The SI units used in turbomachinery are the meter, kilogram, second, and Kelvin. The rest of the units are derived from these basic units. The difference between units and dimensions should be noted. For example, mass is dimension and is measured in units of kilograms. A dimension is therefore a property that can be measured. It is independent of a unit system. A unit assigns number to the dimension while a dimension is a measure of physical quantity. In the English System (ES) of units the primary dimensions are force, length, and time, hence the FLT designation. Force is considered the primary unit in the ES of units and is assigned a nonderived unit. This is a source of confusion and error that necessitates the use of a dimensional constant (g_c) in many equations when solving problems. In ES the unit of g_c is $32.174\frac{lbm \cdot ft}{lbf \cdot s^2}$. Using this constant, the equation of motion can be written as $F = \frac{ma}{g_c} \equiv \frac{lbm \cdot ft \cdot s^{-2}}{lbm \cdot ft \cdot lbf^{-1} \cdot s^{-2}} \equiv [lbf]$, yielding the unit of force to be pound-force. Here the symbol ≡ is used to mean the unit of "something" is equal to. This designation will be used throughout the book. In the SI unit, the unit of g_c is $1\frac{kg \cdot m}{N \cdot s^2}$. However, it is not necessary to use in Newton's equation of motion since force is not a primary unit in the SI system of units.

1.3.1 Primary/Base Dimensions and Units

Primary dimensions are those which are defined as independent ones. They are fundamental and other dimensions can be derived from them. Table 1.1 gives fundamental units and dimensions.

In turbomachinery the most frequently used fundamental dimensions are mass, length, time, and temperature.

1.3.2 Derived Dimensions and Units

Derived/secondary dimensions are those derived from the Table 1.1 seven primary dimensions. For example, in the SI system of units, the dimension of force is a derived unit, which has the dimension of mass multiplied by the dimension of acceleration. The IP dimension is confusing and does not have a systematic and logical base. Units are arbitrarily related to each other. For example: 1 foot = 12 inches, 1 mile = 5280 feet. Table 1.2 gives some derived quantities.

TABLE 1.1
Fundamental Units

Physical Quantity/	SI (MLT)		ES (FLT)		BG (FLT)	
Dimension	**Unit**	**Symbol**	**Unit**	**Symbol**	**unit**	**Symbol**
Mass	kilogram	kg	pound mass	lbm	slug	slug
Length	meter	m	foot	ft	foot	ft
Time	second	s	second	s	second	S
Temperature	Kelvin	K	Rankine	R	degree Rankine	°R
Electric current	ampere	A	ampere	A	ampere	A
Amount of light	candela	c	candela	c	candela	c
Amount of matter	mole	mol	mole	mol	mole	mol

SI, International System of Units; ES, English System of Units; BG, British Gravitational; MLT, mass, length, time; FLT, force, length, and time.

TABLE 1.2
Derived Units

Physical Quantity/ Dimension	SI		IP	
	Unit	Symbol	Unit	Symbol
Velocity	$m•s^{-1}$	v	fts^{-1}	v
Acceleration	$m•s^{-2}$	a	fts^{-2}	A
Force	Newton	N	pound force	lb_f
Stress	$N•m^{-2}$	P_a	psi (pound per square inch); $lb_f ft^{-2}$ (pound force per square feet)	psi or $lb_f•ft^{-2}$
Pressure	$N•m^{-2}$	P_a	psi (pound per square inch); $lb_f ft^{-2}$ (pound force per square feet)	psi or $lb_f•ft^{-2}$
Density	$kg•m^{-3}$	ρ	$lb_m•ft^{-3}$	ρ
Work	Joule (N•m)	J	$lb_f•ft$	W
Energy	Joule (N•m)	J	BTU	Btu
Power	Watt ($N•m•s^{-1}$)	W	BTU per hour	Btu/h
Heat	J		BTU	Btu
Heat capacity	$J•K^{-1}$	C	BTU per hour; horsepower	Btu/F; hp
Specific heat capacity	$J•K^{-1}•kg^{-1}$	c	BTU per pound mass per Fahrenheit	$Btu/lb_m•F$
Dynamic viscosity	$Kg•m^{-1}•s^{-1}$	μ	Pound mass per foot per hour	$lb_m/ft•h$; $lb_f•s/ft^2$
Kinematic viscosity	$m^2•s^{-1}$	ν	Square feet per second	ft^2/s
Volumetric flow rate	$m^3•s^{-1}$	Q	Cubic feet per minute	Cfm
Mass flow rate	$kg•s^{-1}$	$\dot{m}$	Pound mass per hour	$lb_m•h^{-1}$
Specific weight	$kg•m^{-2}•s^{-1}$	g	Pound force per cubic foot	lb_f/ft^3

SI, International System of Units; IP, Imperial Units; BTU, British thermal unit.

1.4 COMMONLY USED QUANTITIES IN TURBOMACHINES

1.4.1 Flow Rate

One of the commonly used quantities to describe a turbomachine performance is flow rate, for example, to find out the flow rate of a pump at a given pressure. Flow rate can be given in terms of volumetric flow rate (volume of water discharged from the outlet area) per unit time or mass flow rate (mass of liquid discharged from the outlet area per unit time). We will use the symbols Q for volumetric flow rate and $\dot{m}$ for mass flow rate in this book.

The relationship between volumetric flow rate and mass flow rate is given by Equation 1.1.

$$\dot{m} = \rho Q \tag{1.1}$$

where
Q is volumetric flow rate in m^3/s, or ft^3/s
$\dot{m}$ is mass flow rate in kg/s, or lbm/s
ρ is density in kg/m^3, lbm/ft^3.

The sum of flow rate and leakage flow, through seals, for example, is generally known as theoretical flow. The International Organization for Standardization (ISO) defines normal flow rate as volumetric flow rate related to conditions at which usual operation is expected. Further, we define the following terms, which are specifically related to pumps.

- Optimal flow rate, Q_{opt}, is flow rate at to the best efficiency point.
- Minimum flow rate, Q_{min}, is the allowable minimum flow rate at which the equipment can operate.
- Minimum continuous thermal flow, $Q_{min,\ therm}$, is the lowest flow at which a pump can operate without its operation being diminished by the temperature rise of the pumped liquid.
- Minimum continuous stable flow (MCSF), $Q_{min,\ stable}$, is the minimum flow rate required for continuous operation. It is important that a minimum flow rate is maintained for proper operation of the equipment. Otherwise, mechanical damage due to overheating, cavitation, and recirculation can occur. An MCSF is the flow below which the pump should not be operated continuously. The primary purpose of an MCSF is to achieve satisfactory bearing and seal life.

Figure 1.4 schematically illustrates the effects of operating pumps away from their best-efficiency point:

1. Temperature rise
2. Vibration, cavitation, noise
3. Minimum acceptable flow
4. Impeller cavitation erosion limit (beyond this limit the impeller will be subject to cavitation erosion)
5. Discharge or suction recirculation point
6. Best-efficiency point flow
7. High-flow cavitation
8. Minimum continuous thermal flow
9. MCSF

1.4.2 Pressure

Pressure is a normal force exerted by a fluid (liquid or gas) per unit area. The unit of pressure is pascal (1 Pa = 1 N/m^2). Other units commonly used are:

1 atm = 101.325 kPa
1 bar = 10^5 Pa = 100 kPa = 760 mm column of mercury at 0°C, with density of 13,595 kg/m^3 under gravitational acceleration $g = 9.807\ m/s^2$).

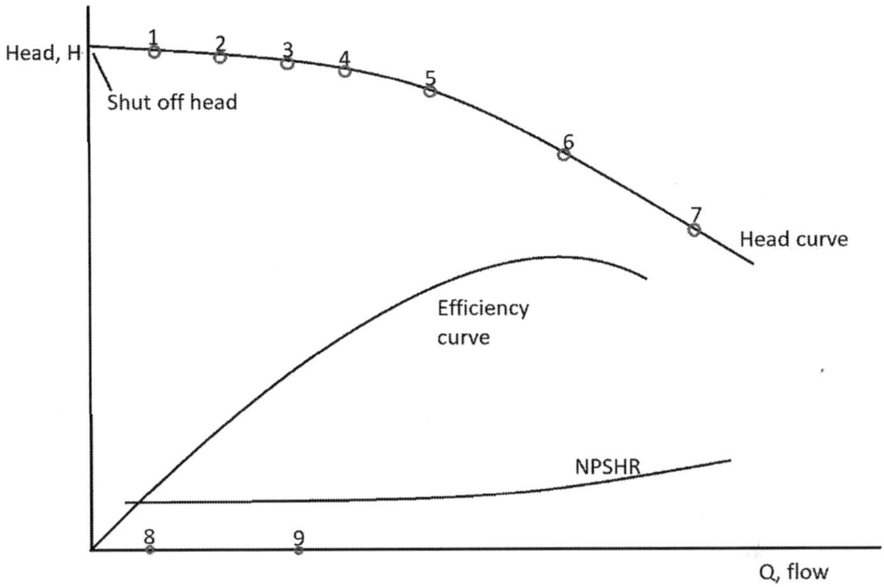

FIGURE 1.4 Effects of operating pumps away from their best-efficiency points.

1 atm = 760 torr (in honor of Torricelli)
1 torr = 133.3 Pa.

In the imperial system, the unit of pressure is pound-force per square inch:

1 psi = 1 lbf/in^2
1 atm = 14.696 psi

The above values indicate pressure measured at sea level. Pressure decreases with increase in altitude. Pressure measurements are further divided into three categories:

1. Absolute pressure: pressure measured relative to absolute-zero pressure
2. Gage pressure: the difference between absolute pressure and atmospheric pressure
3. Vacuum pressure: the difference between atmospheric pressure and absolute pressure. It is noted that pressure below atmospheric pressure is known as vacuum pressure.

Figure 1.5 illustrates these pressures.

Gage and vacuum pressures can be calculated as follows:

$$P_{gage} = P_{abs} - P_{atm} \tag{1.2}$$

$$P_{vac} = P_{atm} - P_{abs} \tag{1.3}$$

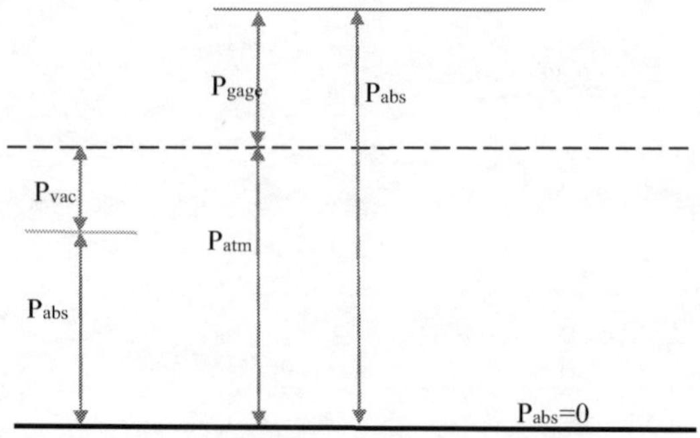

FIGURE 1.5 Absolute, gage, and vacuum pressures.

In IP units psia is used to indicate that the pressure is absolute, and psig is used to indicate that the pressure is gage pressure.

1.4.3 Head and Power

One of the most important quantities in the design of pumps and turbines is head. In case of hydro-turbines, a *head* is described as head extracted from the fluid by the hydro-turbine, $h_{turbine}$.

$$h_{turbine} = \frac{(P_t)_{in}}{\rho g Q} \tag{1.4}$$

where ρ is density of the fluid, g is acceleration due to gravity, Q is discharge (volumetric flow rate) and $(P_t)_{in}$ is waterpower. Waterpower is power extracted from the water. Note that in some literature, water horsepower is a name given to $(P_t)_{in}$ indicating power extracted from the water; the unit does not have to be in horsepower.

An important quantity in every pump design is net head developed by the pump. The total head has a unit of length (ft, m). The total head (manometric head) is a method of expressing the pressure developed by the pump. The head describes pressure rise in terms of the height of a column of liquid. The relationship between the manometric head of the pump, h_{pump}, power output, $P_{p\,out}$, and discharge is given by Equation 1.5.

$$h_{pump} = \frac{(P_p)_{out}}{\rho g Q} \tag{1.5}$$

1.4.3.1 Head of a Pump Between Two Reservoirs

Consider Figure 1.6, in which an axial pump is installed. The pump transfers water from the tank on the left to the tank on the right. The pressure developed by the pump is given by:

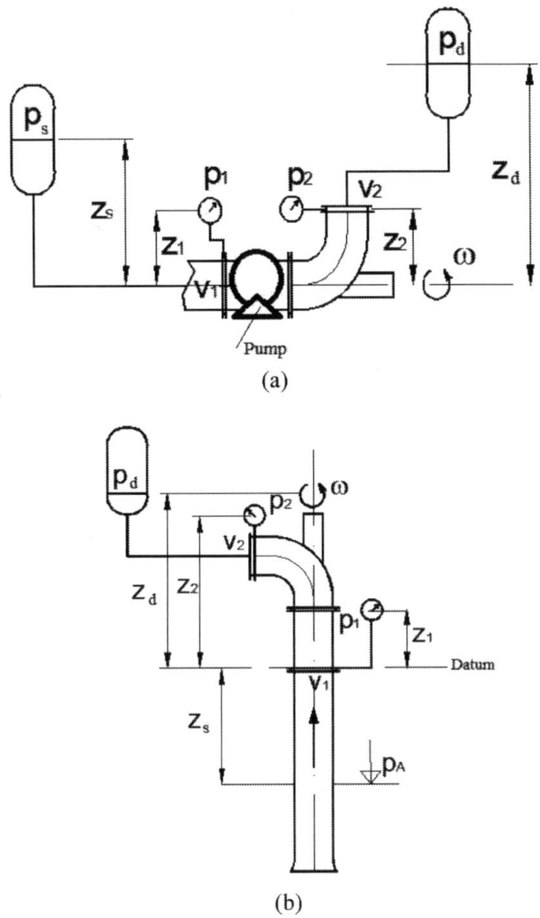

FIGURE 1.6 (a) Horizontal pump layout; (b) vertical pump layout.

$$\Delta p = p_2 - p_1 + \rho g(z_2 - z_1) + \rho\frac{v_2^2 - v_1^2}{2} \tag{1.6}$$

In terms of head

$$H = \frac{p_2 - p_1}{\rho g} + (z_2 - z_1) + \frac{v_2^2 - v_1^2}{2g} \tag{1.7}$$

where p_1 is pressure on the suction side measured at the inlet to the pump
p_2 is pressure in the discharge side measured at the outlet of the pump
v_1 is suction side velocity
v_2 is discharge side velocity
z_1 is suction side elevation measured from the centerline

z_2 is discharge side elevation measured from the centerline
Δp is pressure difference
H is head
ρ is density
g is acceleration due to gravity
z_s is elevation of the water level on the suction side measured from the centerline
z_d is elevation of the water level on the discharge side measured from the centerline
ω is angular velocity of the pump impeller
p_A is atmospheric pressure.

1.4.4 Efficiency

Irreversible losses in pumps and hydro-turbines are given by their efficiencies. The pump converts the mechanical energy through a rotating shaft (hence shaft work) to the fluid by increasing its flow velocity, potential energy (elevation), and pressure. On the other hand, a hydro-turbine extracts energy from the fluid and converts it to mechanical energy through a rotating shaft (hence shaft work). It is during these conversions that irreversible losses in pumps and hydro-turbines occur, laying the foundation for efficiency definition. Pump efficiency is defined as:

$$\eta_{pump} = \frac{\left(P_p\right)_{out}}{\left(P_p\right)_{in}} = \frac{\dot{W}_{pump,u}}{\dot{W}_{pump}} \tag{1.8}$$

where $\dot{W}_{pump,u}$ is useful pumping power and $\dot{W}_{pump}$ is power input to the pump (mechanical energy input).

Turbine efficiency is defined as:

$$\eta_{turbine} = \frac{\left(P_t\right)_{out}}{\left(P_t\right)_{in}} = \frac{\dot{W}_{turbine}}{\dot{W}_{turbine,e}} \tag{1.9}$$

where $\dot{W}_{turbine,e}$ is mechanical power extracted from the water by the turbine and $\dot{W}_{turbine}$ is shaft power of the turbine.

Usually, a pump is packaged with a motor and a turbine comes with a generator. Therefore, it is important to define motor efficiency and generator efficiency. Motor efficiency is defined as follows:

$$\eta_{pump} = \frac{\left(P_m\right)_{out}}{\left(P_m\right)_{in}} \tag{1.10}$$

where $\left(P_m\right)_{out}$ is shaft power output from the motor and $\left(P_m\right)_{in}$ power input to the motor from the source.

The overall efficiency of the pump–motor package can be calculated as follows:

$$\eta_{pump-motor} = \eta_{pump}\eta_{motor} \tag{1.11}$$

Generator efficiency is defined as follows:

$$\eta_{generator} = \frac{\left(P_g\right)_{out}}{\left(P_g\right)_{in}} \tag{1.12}$$

where $\left(P_g\right)_{out}$ is electrical power output from the generator and $\left(P_g\right)_{in}$ shaft power input from the turbine.

The overall efficiency of a turbine–generator package is given by:

$$\eta_{turbine-generator} = \eta_{turbine}\eta_{generator} \tag{1.13}$$

Equations 1.8–1.13 illustrate that the conversion of power by the pump–motor or turbine–generator package is less than 100% due to irreversible losses.

1.4.5 Specific Speed

Pumps are divided into three categories, as indicated in Figure 1.3: radial, mixed-flow, and axial. An engineer is regularly faced with the task of deciding the type of pump (turbomachine) to use for a specific application. One crucial tool available to the engineer is specific speed. Knowing the head, H, flow rate, Q, and rotational speed, N, the engineer can use a nondimensional parameter known as *specific speed.* The specific speed expresses the variation of H, Q, and N which causes similar flow patterns in turbomachines that are geometrically similar. The nondimensional specific speed n_b is given by:

$$n_b = \frac{NQ^{\frac{1}{2}}}{\left(gH\right)^{\frac{3}{4}}} \tag{1.14}$$

where

n_b is specific speed (dimensionless)
N is speed (rad.s^{-1})
g is acceleration due to gravity (ms^{-2})
H is head (m)
Q is flow rate (m^{3}s^{-1}).

Figure 1.7 shows relationship between specific speed and type of pump. It also shows relationship between specific speed and pump performance characteristics.

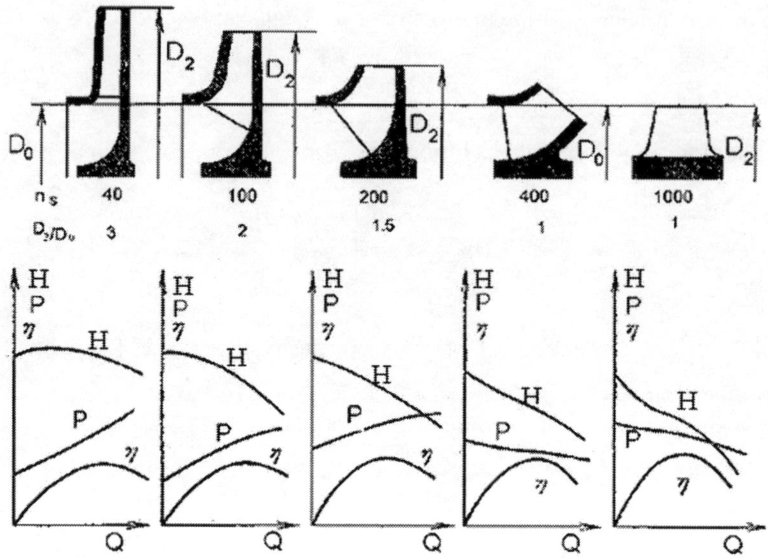

FIGURE 1.7 Relationship between specific speed and pump type and shape of performance characteristics.

Nondimensional specific speed can also be written in terms of N, P, and H, as follows:

$$n_b = N\frac{P^{\frac{1}{2}}}{\rho^{\frac{1}{2}}\left(gH\right)^{\frac{5}{4}}} \tag{1.15}$$

where

n_s is specific speed (dimensionless)
N is speed (s^{-1})
g is acceleration due to gravity (ms^{-2})
H is head (m)
P is power (W)
ρ is density (kgm^{-3}).

Other ways of writing specific speed are listed below.

$$n_b = 3.65N\frac{Q^{\frac{1}{2}}}{H^{\frac{3}{4}}} \tag{1.16}$$

where N is in rpm, Q in m^3/s, and H in m. In some literature it is common to see dimensional specific speed (designated as n_s) in the following form:

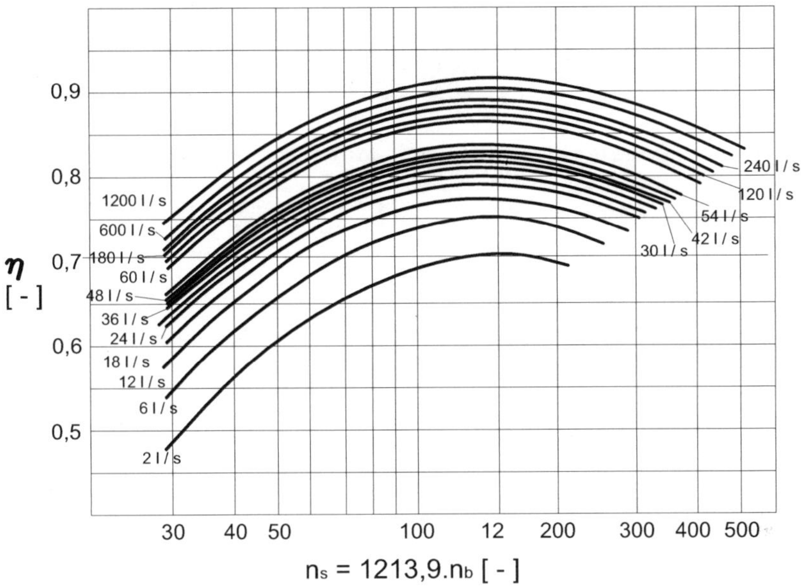

FIGURE 1.8 Maximum achievable efficiency of pumps.

$$n_s = N\frac{Q^{\frac{1}{2}}}{H^{\frac{3}{4}}} \tag{1.17}$$

where N is in rpm, Q in gpm, and H in ft.

Figure 1.8 shows the relationship between maximum achievable efficiency at a given specific speed. It is noted that $n_s = 1213.9n_b$.

1.5 EXAMPLE PROBLEMS

Example 1.1

Derive primary dimensions for the MLT system of units for the following variables:

(a) Density
(b) Force
(c) Pressure
(d) Dynamic viscosity
(e) Heat.

Solution

(a) $\rho = \frac{m}{V} \equiv \frac{kg}{m^3} \equiv \left[ML^{-3}\right]$

where m is mass and V is volume.

(b) $F = ma \equiv kg\frac{\frac{m}{s}}{s} \equiv kg\frac{m}{s^2} \equiv \left[MLT^{-2}\right]$

where m is mass and a is acceleration.

(c) $P = \frac{F}{A} \equiv \frac{N}{m^2} \equiv \frac{kgms^{-2}}{m^2} \equiv kgm^{-1}s^{-2} \equiv \left[ML^{-1}T^{-2}\right]$

where F is force and A is area.

(d) $\mu = \frac{shear\ stress}{velocity\ gradient} \equiv \frac{Pa}{\frac{ms^{-1}}{m}} \equiv \frac{Nm^{-2}}{s^{-1}} \equiv \frac{\frac{kgm}{s^2}m^{-2}}{s^{-1}} \equiv kgm^{-1}s^{-1} \equiv \left[ML^{-1}T^{-1}\right]$

where Pa is Pascal and N is Newton.

(e) $Heat \equiv J \equiv Nm \equiv \frac{kgm}{s^2}m \equiv \left[ML^2T^{-2}\right]$

where J is joule.

Example 1.2

Derive primary dimensions for the FLT system of units for the following variables:

(a) Acceleration
(b) Power
(c) Work
(d) Angular velocity.

Solution

(a) $a = \frac{velocity}{time} \equiv \frac{ft}{s} \equiv \left[LT^{-1}\right]$
(b) Work is force times distance. $W = F \cdot s \equiv lbf \cdot ft \equiv \left[FL\right]$
(c) Power is the time rate of doing work. $P \equiv \frac{lbf \cdot ft}{s} \equiv \left[FLT^{-1}\right]$
(d) Angular velocity is the time rate of angular displacement. $Angular\ velocity = \frac{Angular\ displacement}{time} \equiv s^{-1} \equiv \left[T^{-1}\right]$

EXAMPLE 1.3

What are the primary dimensions for the following variables in the MLT system of units if μ is dynamic viscosity, y is length, u is velocity, ρ is pressure, and r is density?

(a) $\mu\left(\frac{\partial u}{\partial y}\right)$

(b) $\rho\left(\frac{\partial u}{\partial t}\right)$

(c) $\frac{\partial p}{\partial y}$

(d) $\int \frac{\partial^2 u}{\partial t} \partial x$

Solution

(a) $\mu\left(\frac{\partial u}{\partial y}\right) \equiv kgm^{-1}s^{-1}\frac{ms^{-1}}{m} \equiv \left[ML^{-1}T^{-2}\right]$

(b) $\rho\left(\frac{\partial u}{\partial t}\right) \equiv kgm^{-3}\frac{ms^{-1}}{s} \equiv \left[ML^{-2}T^{-2}\right]$

(c) $\frac{\partial p}{\partial y} \equiv \frac{kgm^{-1}s^{-2}}{m} \equiv \left[ML^{-2}T^{-2}\right]$

(d) $\int \frac{\partial^2 u}{\partial t} \partial x \equiv \frac{ms^{-1}}{s} m \equiv \left[ML^{2}T^{-1}\right]$

EXAMPLE 1.4

The lifting force acting on a body in fluid flow is given by: $F_L = \frac{1}{2} C_L \rho v^2 A$

where F_L is the lifting force, C_L is lift coefficient, ρ is density, and A is area. By deriving primary dimensions show that the lift coefficient is indeed dimensionless.

Solution

$$F_L = \frac{1}{2} C_L \rho v^2 A$$

$$C_L = \frac{2F_L}{\rho v^2 A} \equiv \frac{2kgms^{-2}}{kgm^{-3}m^2 s^{-2} m^2} \equiv [2]$$

$$\therefore C_L \text{ is dim } ensionless$$

Example 1.5

The drag force exerted on a spherical object with small Reynolds number can be expressed using the Stokes' law: $F_D = 6\pi\mu R v$, where F_D is the drag force, μ is dynamic viscosity, R is radius of the sphere, and v is the velocity of the sphere. Show that F_D indeed has the unit of force.

Solution

$$F_D = 6\pi\mu R v \equiv 6\pi\left(kgm^{-1}s^{-1}\right)(m)\left(ms^{-1}\right) \equiv kgms^{-2} \equiv \left[MLT^{-2}\right]$$

Example 1.6

What is the absolute pressure at a given point if the gage pressure is measured to be 320 psig?

Solution

$$P_{abs} = P_{atm} + P_{gage}$$
$$P_{abs} = 14.7\,psi + 320\,psi$$
$$P_{abs} = 334.7\,psia$$

1.6 EXERCISE PROBLEMS

1. Derive the primary dimension for the MLT system of units for the following:
   (a) Specific weight
   (b) Power
   (c) Flow rate
   (d) Energy.
2. Show that the equation for the drag force given by $F_D = \frac{1}{2} C_D \rho v^2 A$ is dimensionally homogeneous (i.e., the dimension on the left-hand side is the same as the dimension on the right hand-side).
3. Show that the following equations: (a) Bernoulli equation and (b) normal stress in cylindrical coordinate systems are dimensionally homogeneous.

   (a) $$\frac{p_1}{\rho} + \frac{v_1^2}{2} + gz_1 = \frac{p_2}{\rho} + \frac{v_2^2}{2} + gz_2$$

   $$\sigma_{rr} = -p + 2\mu \frac{\partial v_r}{\partial r}$$

   (b) $$\sigma_{\theta\theta} = -p + 2\mu\left(\frac{1}{r}\frac{\partial v_\theta}{\partial \theta} + \frac{v_r}{r}\right)$$

   $$\sigma_{zz} = -p + 2\mu \frac{\partial v_z}{\partial z}$$

   where σ is stress, μ is dynamic viscosity, v is velocity, and r is radius.

4. If l is length, V is velocity, and v is kinematic viscosity, which of the following combinations give dimensionless quantities?
   (a) $\dfrac{Vl^2}{v}$
   (b) $\dfrac{Vl}{v}$
   (c) $\dfrac{V}{lv}$
   (d) $\dfrac{v}{Vl}$
5. If a pressure loss in a pipe can be expressed by the equation given below, where Δp is pressure loss, V is velocity, l is pipe length, and V is flow velocity, determine the primary dimensions of the constant A.

$$\Delta p = A\left(\frac{L}{\sqrt{D}}\right)\left(V^2\right)$$

1.7 BIBLIOGRAPHY

Çengel, Y. A. and Cimbala, J. M. *Fluid Mechanics: A Fundamental Approach.* New York: McGraw Hill, 2018.

Logan Jr., E. *Handbook of Turbomachinery* (2nd ed.). Boca Raton, FL: Arizona State University/Marcel Dekker, 2003.

Shalaby, A. I. *Fluid Mechanics for Civil and Environmental Engineers.* Boca Raton: CRC Press, 2018.

Stepanoff, A. J. *Centrifugal and Axial Flow Pumps: Theory, Design, and Application.* Malabar, FL: Krieger, 1991.

Turton, R. K. *Principles of Turbomachinery*, The Netherlands: Springer, 1984.

Varchola, M. and Hlobočan, P. *Hydraulic Design of an Axial Machine.* Bratislava, Slovakia: STU Bratislava, 2015.

Varchola, M. and Hlobocan, P. *Hydraulic Design of Centrifugal Pumps.* Bratislava, Slovakia: STU Bratislava, 2016.

2 Scaling Laws and Dimension Analysis

2.1 INTRODUCTION

Due to the presence of a large number of variables, difficulty in relating them analytically, and difficulty in solving the governing fluid flow equations analytically, the design of turbomachines is very complex. A mathematical procedure, known as *dimensional analysis,* is used to obtain the general behavior of turbomachines. Dimensional analysis is a way of streamlining a physical problem by utilizing dimensional homogeneity to reduce the number of relevant variables. In dimensional analysis quantities are expressed in terms of their primary dimensions and grouped to represent some physical situation. Dimensional analysis is a powerful tool that can be used in the absence of sufficient information to set up precise equations describing flow in turbomachines. Dimensional analysis can also be used to predict the performance of a prototype from tests performed on models and to determine the best turbomachine based on the performance. Dimensional analysis results in reduced quantities, thus helping reduce the time and cost involved in experiments on turbomachines. It enables scaling up of results of models to prototypes. Tests made on scale models can be used to assess the performance of a prototype. The scaling laws developed help predict the performance of turbomachines under varying operating conditions. In this book, it is assumed that the reader understands the fundamental concepts of dimensional analysis, scaling laws, and similitude. Therefore, detailed text on fundamental concepts will not be presented here.

2.2 SIMILARITY

Dimensional analysis provides a way to plan and carry out experiments and enables scaling up of model results to prototype. The notion of similarity underpins dimensional analysis. A model and a prototype are hydrodynamically similar if they are geometrically, kinematically, and dynamically similar. These are necessary conditions for hydrodynamic similarity between a model and a prototype.

1. Geometric similarity: the model and the prototype have the same shape, but they are scaled by some constant factor.
2. Kinematic similarity: the velocities of the model and the prototype are proportional at any point of the flow. Flow streamlines must have the same shape. For example, fluid particles in a model and a prototype in translation follow similar paths.
3. Dynamic similarity: the model and prototype flows should have the same force ratios. To the comparable forces in the prototype flow, all forces in

DOI: 10.1201/9781003007142-2

the model flow must scale by the same factor. For example, at equivalent places, the ratio of forces causing fluid acceleration must be identical. As an example, let's consider the one-dimensional Navier–Stokes equation in the z-direction.

$$\frac{\partial w}{\partial t}+w\frac{\partial w}{\partial z}=-\frac{1}{\rho}\frac{\partial p}{\partial z}+v\frac{\partial^2 w}{\partial z^2}-g \tag{2.1}$$

where w is velocity in the z-direction, v is kinematic velocity, g is acceleration due to gravity, ρ is density, and p is pressure. We will nondimensionalize each component in the equation by dividing by some reference value: w_0, p_0, L, t_0

$$w^* = w / w_0,\quad p^* = p / p_0,\quad z^* = z / L,\quad t^* = t / t_0 \tag{2.2}$$

The Navier–Stokes equation in a nondimensional form can be written as:

$$Sr\frac{\partial w^*}{\partial t^*}+w^*\frac{\partial w^*}{\partial z^*}=-Eu\frac{\partial p^*}{\partial z^*}+\frac{1}{\mathrm{Re}}\frac{\partial^2 w^*}{\partial z^{*2}}+\frac{1}{Fr} \tag{2.3}$$

where

$$Sr=\frac{L}{t_0\, v},\quad Eu=\frac{p_0}{\rho v_0},\quad Fr=\frac{g\,L}{v_o^2},\quad \mathrm{Re}=\frac{v_o\,L}{v} \tag{2.4}$$

Sr: Strouhal, EU: Euler, Fr: Froude, and Re: Reynolds are nondimensional numbers.

The nondimensional numbers given in Equation (2.4) are ratios of forces. For dynamic similarity between models and prototypes, such nondimensional force ratios should be the same. Table 2.1 gives some common nondimensional numbers.

where:

g is acceleration of gravity (ms^{-2})
β is coefficient of volume expansion (K^{-1})
ρ is density of fluid (kgm^{-3})
T_s is surface temperature (K)
T_∞ is fluid temperature (K)
l is characteristic length (m)
v is kinematic viscosity (m^2s^{-1})
t is time (s)
v is velocity (ms^{-1})
σ_p is surface tension (Nm^{-1})
p is local pressure (Nm^{-2})
p_v is vapor pressure (Nm^{-2})
E is bulk modulus elasticity (Nm^{-2})
τ_w is shear stress (Nm^{-2}).

TABLE 2.1
Overview of Important Similarity Criteria

Group	Interpretation	Relationship
Grashof number (*Gr*)	Ratio of buoyancy to viscous forces	$Gr = \frac{g\beta(T_s - T_\infty)l^3}{v^2}$
Euler number (*Eu*)	Pressure force / inertia force	$Eu = \frac{p}{\rho \cdot v^2}$
Reynolds number (*Re*)	Inertial force / viscous force	$\text{Re} = \frac{v \cdot l}{\nu}, \text{Re} = \frac{\rho \cdot v \cdot l}{\mu}$
Strouhal number (*Sh*)	Local inertia force / convective inertia force	$Sh = \frac{\varpi}{v}$ ω: frequency of oscillation
Froude number (*Fr*)	Inertial force / gravitational force	$Fr = \frac{v^2}{g \cdot l}$
Mach number (*M*)	Inertia / compressive force	$M = v\sqrt{\frac{\rho}{E}}, M = \frac{v}{c}$
Weber number (*We*)	Inertial force / capillary force	$We = \frac{\rho \cdot l \cdot v^2}{\sigma_p}$
Fourier number (*Fo*)	Ratio of diffusive or conductive transport	$Fo = \frac{v \cdot t}{l^2}$
Cavitation number (s_c)	Ratio of excess local static pressure head to velocity head	$\sigma_c = \frac{2(p - p_v)}{\rho \cdot v^2}$
Cauchy numbe©*C*)	Inertial forces / elastic forces	$C = \frac{\rho \cdot v^2}{E}$
Friction coefficient (f_t)	Pressure force / inertia force of turbulence	$f_t = \frac{\tau_w}{\rho \cdot v^2}$

2.3 RAYLEIGH'S METHOD OF DIMENSIONAL ANALYSIS

Rayleigh method of dimensional analysis, also known as Rayleigh's identical method, expresses a functional relationship of some variables in the form of an exponential equation. It is based on the fundamental principle of dimensional homogeneity of physical variables involved in a problem. The procedures of Rayleigh's method are given below:

1. Identify the dependent variable and express it as a product of all the independent variables raised to an unknown integer exponent.
2. Equate the indices of n fundamental dimensions of the variables involved, to obtain n independent equations.
3. Solve these n equations to obtain the dimensionless groups.

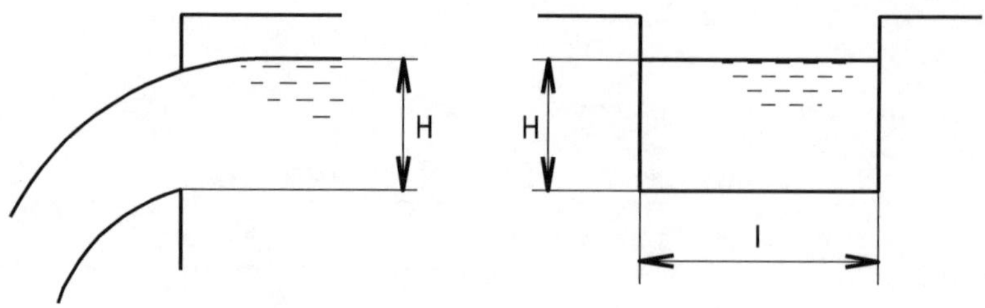

FIGURE 2.1 Flow over a rectangular weir.

We will explain the Rayleigh's method with an example. Let's consider flow through a rectangular weir (Figure 2.1). The volumetric flow rate is given by Equation 2.5.

$$Q = k \cdot l^a \cdot g^b \cdot H^c \tag{2.5}$$

where
- Q is volumetric flow rate
- k is constant
- l is rectangular opening width
- g is acceleration due to gravity
- H is rectangular weir height
- a, b, and c are exponents.

Equation 2.5 can be written in terms of the primary dimensions (mass, length, time (MLT) system) as follows.

$$\frac{L}{T^3} = L^a \cdot \left(\frac{L}{T^2}\right)^b \cdot L^c \Rightarrow L^3 \cdot T^{-1} = L^{a+b+c} \cdot T^{-2b} \tag{2.6}$$

$$Q = k\left[L\right]^a \left[g\right]^b \left[H\right]^c \tag{2.6}$$

$$L^3T^{-1} = \left[L\right]^a \left[g\right]^b \left[H\right]^c \tag{2.7}$$

$$\therefore\ 3 = a + b + c \Rightarrow c = 3 - a - b; \quad b = \frac{1}{2} \tag{2.8}$$

$$Q = k\left[L\right]^a \left[g\right]^{\frac{1}{2}} \left[H\right]^{3-a-\frac{1}{2}} \tag{2.9}$$

$$Q = k \cdot L\sqrt{g}\left[H\right]^{\frac{5}{2}}\left[\frac{L}{H}\right]^a \tag{2.10}$$

$$Q = \left(k \cdot \sqrt{g} \cdot L \cdot H^{\frac{3}{2}}\right) \tag{2.11}$$

This can be verified by directly solving the following equation analytically as follows.

$$Q = \mu \cdot \iint_A \sqrt{2 \cdot g \cdot H} \cdot dA = \mu \cdot \sqrt{2g} \cdot \int_0^H \sqrt{H} \cdot L\, dH = \mu \cdot \sqrt{2} \cdot \frac{2}{3} \cdot L \cdot \sqrt{g} \cdot H^{\frac{3}{2}} \quad (2.12)$$

where μ is rectangular weir coefficient. Note that we have used the fundamental equation $Q = v \cdot A$, and $v = \sqrt{2 \cdot g \cdot H}$ from Torricelli's theorem. It is noted that Rayleigh's method is not always so straightforward, and the number of equations may be less than the number of unknowns. This requires one of the variables to be expressed in terms of the others. It can also be a daunting task if there are more variables.

2.4 BUCKINGHAM PI THEOREM

Buckingham Pi theorem is a formalization of Rayleigh's method of dimensional analysis. In 1915 Buckingham showed that the number of dimensionless groups needed to define a problem is equal to the total number of variables n (viscosity, surface tension, density) minus the primary dimensions m. If there are n variables in a problem and these variables contain m primary dimensions (e.g., mass, length, and time) the equation relating all the variables will have $(n - m)$ dimensionless groups. Buckingham referred to these groups as Π groups. Buckingham Π theorem tells us that if x_1, x_2, $x_3 \ldots x_n$ are n dimensional variables and these variables contain m primary dimensions (e.g., M, L, T), the equation relating the variables will contain $n - m$ dimensionless groups. Buckingham denoted these dimensionless groups as Π_1, Π_2, $\Pi_3 \ldots \Pi_{n-m}$, and the final equation obtained is: $\Pi_1 = \varphi\ (\Pi_2, \Pi_3 \ldots \Pi_{n-m})$. Buckingham Π theorem helps us form non-dimensional numbers such as the ones given in Table 2.1. For example, in piping systems viscous and inertial forces are of importance. Therefore, in the analysis of the flow in the pipeline, it will be modeled in terms of the Reynolds number.

Often properties such as fluid elasticity, surface tension, and shear stress are of importance during the flow. This means that in dimensional analysis of turbomachines, we need to consider geometric, kinematic, dynamic similarities and physical properties of fluids (Table 2.2).

As an example, consider flow in a circular pipe in which the dimensional variables that are physically relevant may be Q, p, n, l, r, g, σ, τ. These can be expressed in a functional relationship as follows:

$$f\left(Q, p, l, \rho, v, g, \sigma, \tau\right) = 0 \quad (2.13)$$

The n number of variables here is 8 and the m of primary dimensions is 3. Then the remaining variables can be expressed as $(n - m)$ dimensionless independent quantities or 5 Π groups, Π_1, Π_2, $\Pi_3 \ldots \Pi_{n\text{-}m}$. The functional relationship can thus be reduced to the much more compact form:

$$\phi\left(\Pi_1, \Pi_2, \Pi_3 \ldots \Pi_{n-m}\right) = 0 \quad \text{or equivalently} \quad \Pi_1 = \phi\left(\Pi_2, \Pi_3 \ldots \Pi_{n-m}\right) \quad (2.14)$$

TABLE 2.2
An Overview of Physical Quantities Important in Turbomachines

Physical Value	Symbol	Size
Length, roughness, diameter	l, k, d	M
Mass	m	kg
Time	t	s
Velocity	c, v, w, u	$m\cdot s^{-1}$
Acceleration	a	$m\cdot s^{-2}$
Angular velocity	ω	s^{-1}
Force	F	[N], $kg\cdot m\cdot s^{-2}$
Gravitational acceleration	g	$m.s^{-2}$
Volumetric flow rate	Q	$m^3\cdot s^{-1}$
Mass flow rate	$\dot{m}$	$kg\cdot s^{-1}$
Pressure	p	[Pa], $kg\cdot m^{-1}\cdot s^{-2}$
Shear stress	τ	[Pa], $kg\cdot m^{-1}\cdot s^{-2}$
Density	ρ	$kg\cdot m^{-3}$
Specific weight	γ	$kg\cdot m^{-2}\ s^{-2}$
Work	W	[N·m], $kg\cdot m^2\cdot s^{-2}$
Dynamic viscosity	μ	[Pa·s], $kg\cdot m^{-1}\cdot s^{-1}$
Kinematic viscosity	ν	$m^2\cdot s^{-1}$
Power	P	[$N\cdot m\cdot s^{-1}$], $kg\cdot m^2\cdot s^{-3}$
Heat flow	q	$kg\cdot m^2\cdot s^{-3}$
Surface tension	σ	[$N\cdot m^{-1}$], $kg\cdot s^{-2}$
Modulus of elasticity in tension	E	[Pa], $kg\cdot m^{-1}\cdot s^{-2}$
Modulus of elasticity in shear	G	[Pa], $kg\cdot m^{-1}\cdot s^{-2}$

2.4.1 Procedure for Buckingham Pi Method

To illustrate the Buckingham π method, we will consider water flow in a pipe. Assuming water is incompressible we will list the number of n variables involved in the problem. Typically, we pick variables which characterize the fluid properties, flow geometry, flow rate, etc. From the geometrical parameters the following can be used: pipe length l, pipe diameter d, and pipe surface roughness k. We assume that the properties of the water are known. It should also be noted that the shear stress is applied and manifests itself in the pressure drop, Δp. With these simplifying assumptions, we are dealing with the following six variables:

$$l, d, \mathrm{v}, \rho, \Delta p, \mu \quad \text{or} \quad \Delta p = f\left(l, d, \mathrm{v}, \rho, \mu\right) \tag{2.15}$$

where μ is dynamic viscosity.

1. List all the n physical quantities or variables involved in the phenomenon. Note their dimensions and the number m of the fundamental dimensions comprised in them, so that there will be $(n - m)$ Π terms.

2. List the dimensions of each variable according to MLT or [force, length, time (FLT).

l	D	v	ρ	Δp	m
[m]	[m]	$[m \cdot s^{-1}]$	$[kg \cdot m^{-3}]$	$[kg \cdot m^{-1} \cdot s^{-2}]$	$[kg \cdot m^{-1} \cdot s^{-1}]$

3. Select m variables out of these to serve as repeating variables with the following guidelines:
   i. These variables themselves should not be dimensionless.
   ii. No two variables should have the same dimensions.
   iii. The entire m fundamental dimensions are included collectively in them.
   iv. The dependent variable is not taken as a repeating variable.

4. Determine the number of $(n - m)$ Π terms.
   $j = m - n = 6 - 3 = 3$

   $$f(\Pi_1, \Pi_2, \Pi_3) = 0$$

5. Form j dimensionless Π groups and check that they are all indeed dimensionless. These variables will be d, v, ρ with unknown exponent and l, Δp, μ with exponent equal to 1 because they are nonrepeating variables.

$$\Pi_1 = d^a \cdot \mathrm{v}^{\mathrm{b}} \cdot \rho^c \cdot \mu$$

$$kg^0 \cdot m^0 \cdot s^0 = [m]^a \cdot \left[m \cdot s^{-1}\right]^b \cdot \left[kg \cdot m^{-3}\right]^c \cdot \left[kg \cdot m^{-1} \cdot s^{-1}\right]$$

$$0 = c + 1 \qquad \Rightarrow c = -1$$
$$0 = a + b - 3c - 1 \quad \Rightarrow a = -1$$
$$0 = -b - 1 \qquad \Rightarrow b = -1$$

$$\therefore \Pi_1 = d^{-1} \cdot \mathrm{v}^{-1} \cdot \rho^{-1} \cdot \mu = \frac{v}{\mathrm{v} \cdot d} = \frac{1}{\mathrm{Re}}$$

$$\Pi_2 = d^a \cdot \mathrm{v}^{\mathrm{b}} \cdot \rho^c \cdot \Delta p$$

$$kg^0 \cdot m^0 \cdot s^0 = [m]^a \cdot \left[m.s^{-1}\right]^b \cdot \left[kg \cdot m^{-3}\right]^c \cdot \left[kg \cdot m^{-1} \cdot s^{-2}\right]$$

$$0 = c + 1 \qquad \Rightarrow c = -1$$
$$0 = a + b - 3c - 1 \quad \Rightarrow a = 0$$
$$0 = -b - 2 \qquad \Rightarrow b = -2$$

$$\therefore \Pi_2 = d^0 \cdot \mathrm{v}^{-2} \cdot \rho^{-1} \cdot \Delta p = \frac{\Delta p}{\rho.\mathrm{v}^2} = Eu$$

and finally

$$\Pi_3 = d^a \cdot \mathrm{v}^{\mathrm{b}} \cdot \rho^c \cdot l$$

$$kg^0 \cdot m^0 \cdot s^0 = [m]^a \cdot \left[m.s^{-1}\right]^b \cdot \left[kg \cdot m^{-3}\right]^c \cdot [m]$$
$$0 = c \qquad \Rightarrow c = 0$$
$$0 = a + b - 3c + 1 \Rightarrow a = -1$$
$$0 = -b \qquad \Rightarrow b = 0$$

$$\therefore \Pi_3 = d^a \cdot \mathrm{v}^{\mathrm{b}} \cdot \rho^c \cdot l = d^{-1} \cdot \mathrm{v}^0 \cdot \rho^0 \cdot l = \frac{l}{d}$$

6. Express the result in functional form as

$$f\left(\frac{l}{d}, Eu, \frac{1}{\mathrm{Re}}\right) = 0 \tag{2.16}$$

Let's consider Π_2, $\frac{\Delta p}{\rho v^2} = Eu$. This can be expressed in terms of the Π_1 and Π_3 as follows, where *f* stands for function.

$$\frac{\Delta p}{\rho v^2} = \frac{l}{d} f\left(\frac{1}{\mathrm{Re}}\right) \tag{2.17}$$

Or by substituting $\lambda = 2f$

$$\Delta p = \lambda \frac{l}{d} \frac{v^2}{2} \tag{2.18}$$

where $\lambda = f(Re)$.

Equation (2.18) describes pressure loss in pipe flow. If we include pipe roughness k, $\lambda = f\left(\frac{k}{d}, \mathrm{Re}\right)$ and Equation (2.17) will have the following form:

$$\frac{\Delta p}{\rho v^2} = \frac{l}{d} f\left(\frac{k}{d}, \mathrm{Re}\right) \tag{2.19}$$

2.5 PRINCIPLES OF SIMILARITY

Similarity principles are an excellent tool for grouping turbomachines into families. Turbomachine classification aids designers in creating similar turbomachines using the same set of data. For example, it can be used to determine which pump is best for a specific application. For example, the performance of a turbomachine can be expressed in terms of the control variables such as flow rate, geometric variables, and fluid properties, as follows:

$$f\left(Q, H, N, g, \rho, v, \sigma, l_1, l_2, l_3, \ldots\right) = 0 \tag{2.20}$$

where l_1, l_2, and l_3 are geometric variables, ρ, n, σ are fluid properties which influence the flow process, and Q, H, and n (rotational speed) are known as control variables. The flow rate, the head (pressure), and rotational speed can be adjusted as desired, hence they are known as control variables.

For geometrically similar turbomachines, we can apply only one dimension, known as the characteristic dimension. The characteristic dimension for pumps and turbines is the outer diameter of the impeller D_2. Equation 2.20 can then be rewritten as:

$$f\left(Q, H, N, g, \rho, v, \sigma, D_2\right) = 0 \tag{2.21}$$

All the quantities described in Equation 2.21 can be described by three fundamental dimensions, i.e., mass in kg, length in m, and time, t, in seconds.

2.5.1 Specific Speed

A dimensionless term of great importance that may be obtained by manipulating the discharge and head coefficients is the specific speed. This equation can be obtained by dimensional analysis. Starting with Equation 2.21, and using the procedure listed in Section 2.4.1, we get:

1. List all the n physical quantities or variables involved in the phenomenon. Note their dimensions and the number m of the fundamental dimensions comprised in them, so that there will be $(n - m)$ Π terms.
2. List the dimensions of each variable according to MLT or FLT.

Q	H	N	ρ	g	σ	D_2	ν
$[m^3/s]$	$[m]$	$[s^{-1}]$	$[kg \cdot m^{-3}]$	$[m^{-1} \cdot s^{-2}]$	$[kg \cdot s^{-2}]$	$[m]$	$[m^2 \cdot s^{-1}]$

3. Select m repeating variables out of these. Taking kg, m, and σ as repeating variables, we will have
   the number of $(n - m)$ Π terms.
   $j = m - n = 8 - 3 = 5$.
   $f\left(\Pi_1, \Pi_2, \Pi_3, \Pi_4\right) = 0$. Note that because H and D_2 have the same dimensions, the number of independent groups will reduce to 4.
4. Form j dimensionless Π groups and check that they are all indeed dimensionless.

$$\Pi_1 = Q^a \cdot (\text{gH})^{\text{b}} \cdot \rho^c \cdot N$$

$$kg^0 \cdot m^0 \cdot s^0 = \left[m^3 s^{-1}\right]^a \cdot \left[m^2 \cdot s^{-2}\right]^b \cdot \left[kg \cdot m^{-3}\right]^c \cdot \left[s^{-1}\right]$$

$$\begin{aligned} & && \Rightarrow c = 0 \\ 0 &= c && \\ 0 &= 3a + 2b - 3c && \Rightarrow a = \frac{1}{2} \\ 0 &= -a - 2b - 1 && \\ & && \Rightarrow b = -\frac{3}{4} \end{aligned}$$

$$\therefore \Pi_1 = Q^{\frac{1}{2}} \cdot (gH)^{\frac{-3}{4}} \cdot \rho^0 \cdot N = N\frac{\sqrt{Q}}{(gH)^{\frac{3}{4}}}$$

This nondimensional number is known as specific speed.

$$n_b = n\frac{\sqrt{Q}}{(gH)^{\frac{3}{4}}} \tag{2.22}$$

Specific speed is very important and used as a guide to the selection of type of pump or turbine required for a particular role. The physical meaning of specific speed is the rotational speed needed to discharge 1 unit of flow against 1 unit of head. Figure 2.2 illustrates the relationship and type of pump. The figure shows that axial-flow pumps have higher specific speeds compared to, for example, radial-flow pumps.

In a similar manner we can drive the remaining three Π parameters.

$$\Pi_2 = Q^a \cdot (\mathrm{N})^{\mathrm{b}} \cdot \rho^c \cdot D_2 \rightarrow \Pi_2 = \frac{Q}{ND_2} \tag{2.23}$$

$$\Pi_3 = (gH)^a \cdot (\mathrm{N})^{\mathrm{b}} \cdot \rho^c \cdot D_2 \rightarrow \Pi_3 = \frac{D_2\mathrm{N}}{\sqrt{(gH)}} \tag{2.24}$$

$$\Pi_4 = Q^a (gH)^b \cdot \rho^c \cdot D_2 \rightarrow \Pi_4 = \frac{\sqrt{gH}D_2^2}{Q} \tag{2.25}$$

These three equations lay the foundation for the affinity laws (scaling laws) for model and prototype, written as follows, where subscript m stands for model and subscript p stands for prototype.

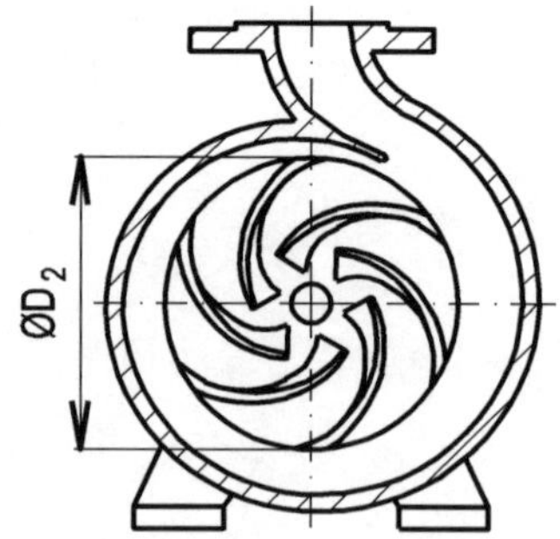

FIGURE 2.2 Cross-sectional view of a pump.

$$\frac{Q_m}{N_m D_{2m}^3} = \frac{Q_p}{N_p D_{2p}^3} \rightarrow \frac{Q_m}{Q_p} = \frac{N_m}{N_p}\left(\frac{D_{2m}}{D_{2p}}\right)^3 \tag{2.26}$$

$$\frac{N_m D_{2m}}{\sqrt{(gH)_m}} = \frac{N_p D_{2p}}{\sqrt{(gH)_p}} \rightarrow \frac{(gH)_m}{(gH)_p} = \left(\frac{D_{2m}}{D_{2p}}\right)^2 \left(\frac{N_m}{N_p}\right)^2 \tag{2.27}$$

$$\frac{\sqrt{(gH)_m} D_{2m}^2}{Q_m} = \frac{\sqrt{(gH)_p} D_{2p}^2}{Q_p} \rightarrow \frac{D_{2m}}{D_{2p}} = \sqrt{\frac{Q_{2m}}{Q_{2p}}}\left(\frac{gH_p}{gH_m}\right)^{\frac{1}{4}} \tag{2.28}$$

Equations 2.26 and 2.27, combined with the power equation $P = \rho Q g H$, give the following affinity equation for a power relationship between a model and prototype:

$$\frac{P_m}{P_p} = \left(\frac{D_{2m}}{D_{2p}}\right)^5 \left(\frac{\rho_m}{\rho_p}\right) \tag{2.29}$$

2.6 EXAMPLE PROBLEMS

EXAMPLE 2.1

A hydrodynamic pump model has the following parameters: flow rate: $Q = 20\,\text{l/s}$ and head = 20.4 m at speed of $N = 24.16\ \text{s}^{-1}$. The impeller outer diameter $D_2 = 300$ mm. Assuming geometrically similar pumps, calculate the flow rate and the head of a pump with outer impeller diameter $D_2 = 2.8$ m at speed of $N = 5\ \text{s}^{-1}$, if it is operated with the same fluid, assuming the efficiency of the model and the prototype are the same.

Solution

Given: Q_m=20 l/s, H_m=20.4 m, n_m=24.16 s^{-1}, D_{2m} = 300 mm = 3 m, D_{2p} = 2.8 m, n_p = 5 s^{-1}

Find: Q_p and H_p.

We will use Equation (2.26) to obtain Q_p:

$$Q_p = Q_m \frac{N_p}{N_m}\left(\frac{D_{2p}}{D_{2m}}\right)^3 = 20ls^{-1}\left(\frac{5s^{-1}}{24.16s^{-1}}\right)\left(\frac{2.8m}{0.3m}\right)^3 = 3365.22ls^{-1}$$

We will use Equation (2.28) to find H_p.

$$H_p = H_m\left(\frac{D_{2p}}{D_{2m}}\right)^2\left(\frac{N_p}{N_m}\right)^2 = 20.4m\left(\frac{2.8m}{0.3m}\right)^2\left(\frac{5s^{-1}}{24.16s^{-1}}\right)^2 = 76.11m$$

Example 2.2

A hydro-turbine model was designed and tested. The model-to-prototype ratio is 1:16. A laboratory test was conducted on the model and complete similarity was maintained. Determine:

(a) the ratio of the circumferential velocities of the model and the prototype, if the circumferential velocity of the prototype is $v = 16\ m/s$
(b) ratio of forces on the blades of the model and the prototype F_{zm}/F_{zp}, assuming the same fluid is used in the model and the part
(c) power ratio of model and work and the prototype, P_m / P_p
(d) ratio of model speed to prototype n_m/n_p, if model speed is $n_m = 600/\text{min}$.

Solution

Given: model-to-prototype ratio (i.e., of the characteristic length) is 1:16. Circumferential velocity of the prototype is 16 m/s.

(a) because velocity and characteristic length ratios are given, we will use the Froude number to solve the problem:

$$Fr_m = Fr_p \rightarrow \frac{v_m^2}{g\,L_m} = \frac{v_p^2}{g\,L_p}$$

$$\frac{v_m}{v_p} = \sqrt{\frac{L_m}{L_p}} = \frac{1}{\sqrt{16}} = \frac{1}{4}$$

$$v_m = v_p \cdot \frac{1}{4} = 16 \cdot \frac{1}{4} = 4\,m/s$$

(b) One of the two primary hydrodynamic forces at work in hydro-turbine rotors is the lift force, which acts perpendicularly to the direction of water flow. This force is given by:

$$F_z = v_z \cdot \rho \cdot \frac{v^2}{2} \cdot A$$

where

F_z is the lift force
ρ is the density of the fluid
v is the flow velocity
A is the area of the turbine blade.

This equation can be used to solve the ratio of forces on the blades of the model and the prototype F_{zm}/F_{zp}.

$$F_{zm} = v_{zm} \cdot \rho_m \cdot \frac{v_m^2}{2} \cdot A_m$$

$F_{zp} = v_{zp} \cdot \rho_p \cdot \frac{c_p^2}{2} \cdot A_p$; $\rho_m = \rho_p$ and representing area in terms of the characteristic length, L: $\frac{A_m}{A_p} = \frac{L_m^2}{L_p^2}$.

$$\frac{F_{zm}}{F_{zp}} = \frac{v_m^2}{v_p^2} \cdot \frac{L_m^2}{L_p^2}$$

Also, as shown in step (a) above: $\frac{v_m^2}{v_p^2} = \frac{L_m}{L}$. Therefore,

$$\frac{F_{zm}}{F_{zp}} = \frac{L_m}{L_p} \cdot \frac{L_m^2}{L_p^2} = \left(\frac{L_m}{L_p}\right)^3 = \left(\frac{1}{16}\right)^3 = \frac{1}{4096}.$$

(c) Since power is described as $P = F \cdot v$, we can write:

$$\frac{P_m}{P_p} = \frac{F_{zm} \cdot v_m}{F_{zp} \cdot v_p} = \left(\frac{L_m}{L_p}\right)^{7/2} = \frac{1}{16384}$$

Note that the power produced by the model is 1/16,384 times the power produced by the prototype.

(d) The model and prototype turbine speed can be obtained using Strouhal's number, $Sh = \frac{v}{N \cdot l}$.

$$\frac{v_m}{N_m \cdot L_m} = \frac{v_p}{N_p \cdot L_p}$$

$$\frac{N_m}{N_p} = \frac{v_m \cdot L_p}{v_p \cdot L_m} = \left(\frac{4m/s}{16m/s}\right)\left(\frac{16}{1}\right) = 4 \rightarrow N_p = \frac{N_m}{4} = \frac{600rpm}{4} = 150rpm$$

Example 2.3

A throttling device, as shown Figure 2.3, was used to equalize flow in the cooling towers of a nuclear power plant. To regulate the flow, it is necessary to know the dependence of the damper flap rotation on the flow resistance coefficient. Find the relationships between the flow rate and frictional losses between the model and the prototype. Calculate the flow rate of the model if the prototype flow rate is 10 m^3/s. The prototype pipe diameter is d_p = 2200 mm and the model pipe diameter is d_m = 200 mm.

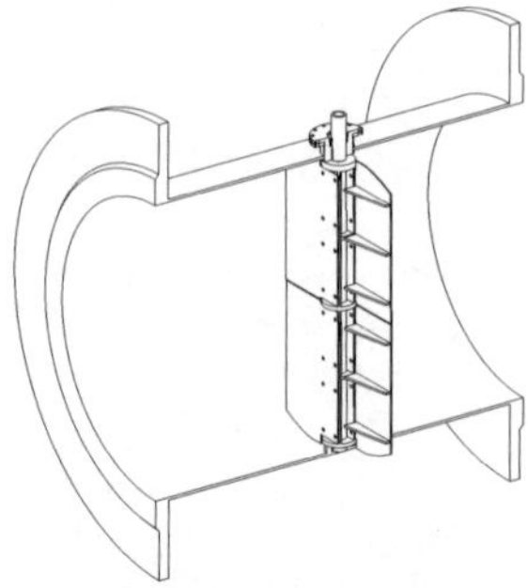

FIGURE 2.3 View of a throttling device.

Solution

Given: $d_p = 2200$ mm, $d_m = 200$ mm. The dependence of the damper rotation on the flow resistance coefficient must be known.

We are asked to model frictional losses as a function of damper flap rotation. This can be done using the Reynolds number, which will be the same for the model and the prototype.

$$\text{Re} = \frac{vd}{\nu} = \frac{4Qd}{\nu\pi d^2} = \frac{4}{\pi}\frac{Q}{d\,\nu}$$

$$\frac{Q_p}{d_p} = \frac{Q_m}{d_m} \Rightarrow Q_m = Q_p\frac{d_m}{d_p} = \frac{1}{11}Q_p$$

$$Q_m = \frac{\text{Re}\cdot d_m \cdot \nu\cdot \pi}{4} = \frac{(578{,}000)\cdot(0.2m)\cdot\left(1\cdot 10^{-6}\,m^2/s\right)\cdot \pi}{4} = 0.09\,m^3/s$$

Example 2.4

A pump fitted with an impeller having 0.15 m diameter is running at a speed of 1500 rpm, delivering 5 m^3/s and at a head of 15 m. Next, a new impeller with a diameter of 0.1 m is used, this time running at a speed of 2200 rpm. Determine the head and the volumetric flow rate of the pump delivered with the newly fitted impeller.

Solution

Given

$$Q_1 = 5\frac{m3}{s} \quad N_1 = 1500rpm \quad N_2 = 2200rpm \quad D_1 = 0.15m \quad D_2 = 0.1m \quad H_1 = 15m$$

$$\frac{H_1}{\left(N_1 D_1^2\right)^2} = \frac{H_2}{\left(N_2 D_2^2\right)^2}$$

$$H_2 = \frac{Q_1 \left(N_2 D_2^2\right)^2}{\left(N_1 D_1^2\right)^2} = 14.341m$$

$$\frac{Q_1}{N_1 D_1^3} = \frac{Q_2}{N_2 D_2^3}$$

$$Q_2 = \frac{Q_1 N_2 D_2^3}{N_1 D_1^3} = 2.173 \frac{m^3}{s}$$

2.7 EXERCISE PROBLEMS

Problem 2.1

A pump is operating at a head of 32 m and discharges 3 m^3/s of water when rotating at 1200 rpm. Its impeller diameter is 1.2 m. A second pump, which is geometrically similar to the first one and has an impeller diameter of 1 m, is operating at 800 rpm. Determine the head and discharge of the second pump. Verify that the specific speeds of the two pumps are same.

Problem 2.2

Calculate the dimensionless specific speed of a turbine with the following parameters:

$Q = 0.15$ m^3/s, $H = 20$ m, $N = 900$ rpm.

Problem 2.3

A turbine operates at a head of 30 m and 260 rpm, and has a flow rate of 7 m^3/s. If the same turbine operates at a head of 15 m, calculate the speed, flow rate, and brake horsepower of the turbine.

Problem 2.4

A hydraulic turbine produces 20 MW power under a head of 40 m and speed of 100 rpm. A geometrically similar model turbine produces 40 kW under a head of 6 m. Determine the model flow rate, speed, and scale ratio.

Problem 2.5

A fan develops pressure head of 200 mm water gauge at a flow rate of 5 m^3/s. The fan runs at a rotational speed of 2000 rpm. Another larger, geometrically similar fan runs at a rotational speed of 1800 rpm. Find the volumetric flow rate of the second larger fan if it operates a pressure head of 200 mm water gauge.

Problem 2.6

A pump with 0.2 m impeller diameter is running at a speed of 2000 rpm; it delivers 5.5 m^3/s and has a head of 18 m. Next, a new impeller with a diameter of 0.15 m is used while running at a speed of 2200 rpm. Determine the head and the volumetric flow rate of the pump fitted with the newly impeller.

Problem 2.7

A centrifugal pump with an impeller diameter of 0.3 m runs at 1650 rpm. A geometrically similar pump with 0.5 m impeller diameter has a rotational speed of 1050 rpm. The head developed by the larger pump is 30 m, while delivering a volumetric flow rate of 0.3 m^3/s. Determine the discharge and the head of the smaller pump at the best-efficiency point. Also determine the ratio of power (P_2/P_1). Verify that the nondimensional specific speed for these geometrically similar pumps is the same.

Problem 2.8

An aircraft's flight speed is 200 m/s, when flying in air having density and dynamic viscosity of 1.22 kg/m^3 and 1.8×10^{-5} kg/m s, respectively. The local speed of sound is 300 m/s. An experimental wind tunnel which allows for the airplane model to be tested at sonic velocity of 350 m/s, density of air 3 kg/m^3, and dynamic viscosity of 1.22×10^{-5} kg/m s is constructed.

Assume full dynamic similarity is maintained and determine the flow velocity and scale of the model.

2.8 BIBLIOGRAPHY

Eck, B. *Technical Fluid Mechanics (in German): Volume 1: Basics*. Berlin: Springer-Verlag, 1988.

Hauke, G. "An introduction to fluid mechanics and transport phenomena," *Fluid Mechanics and its Applications*, The Netherlands: Springer, 2008.

Idelchik, I. E. "Handbook of hydraulic resistance, 2nd edition," *J. Press. Vessel Technol.*, 1987, 109(2).

Spurk, J. H. *Dimension Analysis in Fluid Mechanics*. Berlin: Springer, 1992.

Varchola, M. and Knížat, B. *Fluid Mechanics – Solved Problems (Part II) (in Slovak)*. Bratislavia, Slovakia: STU Bratislava, 2010.

White, F. M. *Fluid Mechanics,* 7th edition. New York: McGraw Hill, 2011.

3 Centrifugal and Mixed-Flow Pumps

3.1 INTRODUCTION

A pump is a machine that imparts energy to a fluid. A centrifugal pump imparts energy to a fluid through rotation of an impeller. The centrifugal pump increases the pressure of the fluid, transports the fluid, and lifts it from a lower to a higher elevation by a centrifugal action. The rotational motion of the impeller increases the momentum of the liquid that enters axially and leaves the pump radially. The impeller is the rotating part while the diffuser is stationary. The accelerating fluid in the blade passages is finally pushed out of the impeller radially. In this way, both static pressure and velocity are increased. The three important parts of a centrifugal pump are: (1) the impeller (shown in Figure 3.1); (2) the diffuser; and (3) volute casing. The stationary volute casing converts the kinetic energy of the fluid, into pressure energy. This happens as the fluid passes through the stationary blade passages which have an increasing cross-sectional area. Finally, the fluid moves from the diffuser blades into the volute casing. The volute casing collects the fluid and transports it to the pump outlet. The volute casing also has a gradually increasing cross-sectional area towards the outlet of the pump for better conversion to pressure energy.

3.2 COORDINATE SYSTEMS USED IN TURBOMACHINERY FLOWS

Cylindrical coordinate systems shown in Figure 3.2 are used in turbomachinery flows. The meridional plane is formed by the radial and axial directions. The tangential velocity component is normal to the meridional plane. In this coordinate system the velocity components are defined as shown in Figure 3.2, showing the axial, radial, and tangential velocity components C_a, C_r, and C_u respectively. The absolute velocity C at any point in the flow is the vector sum of the relative, and tangential velocity components W and U, respectively. The axial, radial, and tangential velocity components of the relative velocity W are W_a, W_r, and W_u, respectively.

3.3 ENERGY TRANSFER IN TURBOMACHINES

Turbomachines are devices in which mechanical energy is transferred from the turbomachine rotor to the fluid or from the fluid to the turbomachine rotor. The physical principle of a turbomachine is based on the basic laws of fluid mechanics, i.e., on the law of conservation of mass, momentum, the first and second laws of thermodynamics. In this book, we will assume that liquid (water) is incompressible, i.e., the density of liquid will be considered constant. This assumption is valid and

DOI: 10.1201/9781003007142-3

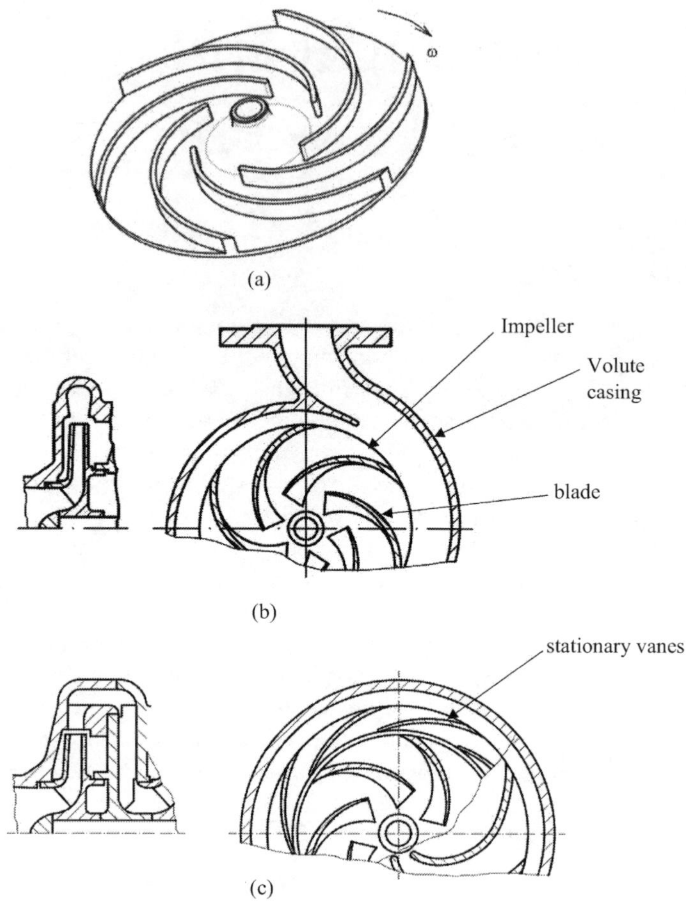

FIGURE 3.1 Centrifugal pump impeller with backward-curved blades. (a) Impeller; (b) impeller in a volute casing; (c) impeller with stationary diffuser.

does cause an insignificant error in calculations unless the pressure in a particular situation is great. With this assumption it is possible to derive theoretical turbomachinery equations starting from the incompressible continuity Equation (3.1), and incompressible Navier–Stokes Equation (3.2).

$$\overline{\nabla}\cdot\overline{V}=0 \tag{3.1}$$

$$\rho\frac{D\overline{V}}{Dt}=-\overline{\nabla}P+\rho\overline{g}+\mu\nabla^{2}\overline{V} \tag{3.2}$$

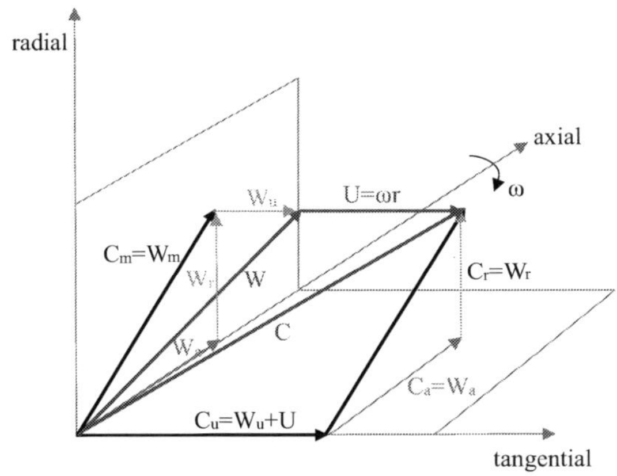

FIGURE 3.2 Velocity components in cylindrical coordinate systems

where:

$\frac{D\bar{V}}{Dt}$ is material (substantial derivative)
$\rho\bar{g}$ is body force
μ is viscosity.

For steady, inviscid, incompressible flow Equation 3.2 can be written as follows:

$$\bar{V}\cdot\bar{\nabla}\bar{V} = \rho\bar{g} - \frac{1}{\rho}\bar{\nabla}P \tag{3.3}$$

Equation 3.3 can be rewritten as follows using vector identity:

$$\bar{\nabla}\left(\frac{V^2}{2}\right) - \bar{V}\times\left(\bar{\nabla}\times\bar{V}\right) = \rho\bar{g} - \frac{1}{\rho}\bar{\nabla}P \tag{3.4}$$

Neglecting body forces, Equation (3.4) becomes:

$$\bar{\nabla}\left(\frac{V^2}{2} + \frac{P}{\rho}\right) = \bar{V}\times\left(\bar{\nabla}\times\bar{V}\right) \tag{3.5}$$

For irrotational flow, the vorticity vector, $\zeta = \left(\bar{\nabla}\times\bar{V}\right) = 0$. Hence:

$$\bar{\nabla}\left(\frac{V^2}{2} + \frac{P}{\rho}\right) = 0 \tag{3.6}$$

Using velocity potential function ϕ for irrotational flow, we can write $\nabla^2\phi = 0$ where ∇^2 is Laplace operator. Assuming that $\frac{\partial}{\partial z} = 0$, in cylindrical coordinate system, the rate of rotation vector (angular velocity vector), which is equal to half of the vorticity vector, can be written as:

$$\bar{\varpi}_z = \frac{1}{2r}\left(\frac{\partial(c_u r)}{\partial r} - \frac{\partial c_r}{\partial \phi}\right) = 0 \tag{3.7}$$

The above equation describes two-dimensional potential, inviscid flow in a pump. This equation is also a necessary condition for potential flow, but not sufficient.

Rewriting Equation (3.4) for axisymmetric flow, i.e., $\frac{\partial}{\partial \phi} = 0$ a $\frac{\partial}{\partial z} = 0$, vorticity vector in cylindrical coordinates, in terms of relative velocity w, we get:

$$w_r \frac{\partial w_r}{\partial r} - \frac{w_u^2}{r} - \omega^2 r - 2\omega w_u = -\frac{1}{\rho}\frac{\partial p}{\partial r} \tag{3.8}$$

$$w_r \frac{\partial w_u}{\partial r} + \frac{w_u w_r}{r} + 2\omega w_r = 0 \tag{3.9}$$

$$w_r \frac{\partial w_z}{\partial r} = 0 \tag{3.10}$$

where w_r is the relative velocity component in the radial direction and w_u is the tangential component of the relative velocity. The equation applies to fluid flow in the impeller channel with an infinite number of infinitely thin plane-curved vanes. Multiplying Equation 3.9 by $-\frac{w_u}{w_r}$ and modifying we get:

$$-\frac{w_u^2}{r} - 2\omega w_u = w_u \frac{\partial w_u}{\partial r} \tag{3.11}$$

Substituting Equation 3.11 for Equation 3.8 and modifying, we get:

$$\frac{1}{\rho}\frac{\partial p}{\partial r} = \omega^2 r - w\frac{\partial w}{\partial r} \tag{3.12}$$

Multiplying Equation 3.12 by ∂r and subsequent integration will yield:

$$\frac{p_2 - p_1}{\rho} = \frac{u_2^2 - u_1^2}{2} + \frac{w_1^2 - w_2^2}{2} \tag{3.13}$$

Equation (3.13) gives the pressure rise across the rotor. In addition to the pressure, the impeller must supply the fluid with the energy needed to increase its kinetic energy. If the fluid's inlet absolute velocity is given by c_1 and exit absolute velocity is given by c_2, then the change in kinetic energy per unit mass of the fluid is $\frac{c_2^2}{2} - \frac{c_1^2}{2}$. The resulting relationship for the theoretical energy transfer per unit mass of fluid in a pump with an infinite number of infinitely thin impeller blades is given by:

$$Y_{t\infty} = \frac{p_2 - p_1}{\rho} + \frac{c_2^2 - c_1^2}{2} = \frac{u_2^2 - u_1^2}{2} + \frac{w_1^2 - w_2^2}{2} + \frac{c_2^2 - c_1^2}{2} \tag{3.14}$$

This is the ideal energy transfer per unit mass for perfect guidance by the impeller blades.

3.4 VELOCITY TRIANGLES

Consider a Cartesian coordinate system consisting of an axial, radial, and tangential components. Consider the impeller rotating with angular velocity ω, as shown in Figure 3.4. The tangential velocity, i.e., the velocity vector in the frame of motion, is denoted by u, the absolute velocity vector is denoted by c, and the relative velocity vector, i.e., the velocity vector in the frame of motion, is denoted by w. The absolute velocity is the sum of the tangential and relative velocity. Figures 3.3 and 3.4 show inlet and outlet velocity triangles

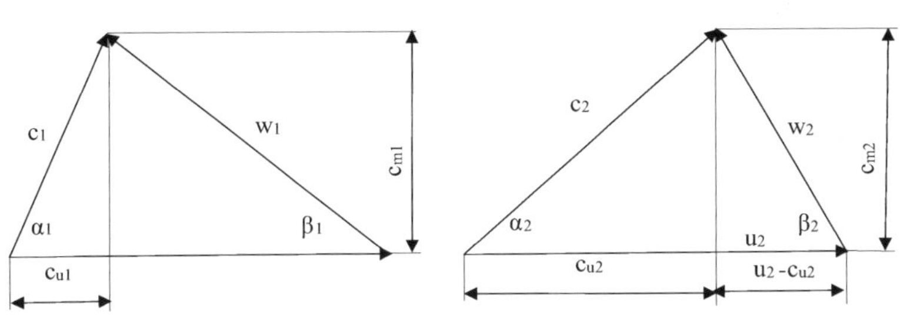

FIGURE 3.3 General case of inlet (a) and outlet (b) velocity triangles.

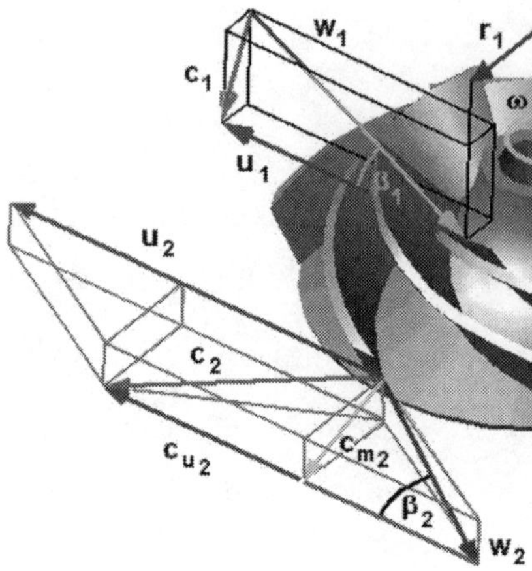

FIGURE 3.4 Inlet and outlet velocity triangles

where:

c is absolute velocity
u is tangential velocity
w is relative velocity
subscripts:
1 is inlet
2 is outlet
u is tangential
m is meridional
α is angle between absolute velocity and tangential velocity components
β is angle between relative velocity and tangential velocity components.

3.5 VELOCITY COMPONENTS FOR DIFFERENT IMPELLER BLADE ORIENTATION

Impeller blade orientation can be:

1. radial ($\beta_2 = 90°$)
2. forward-curved ($\beta_2 > 90°$)
3. backward-curved ($\beta_2 < 90°$).

Radial blades (Figure 3.5a) are easy to design and construct and are mainly used in high-speed compressors. Figure 3.5b shows the velocity triangles for impellers with forward-curved blades at the outlet. Impellers with forward-curved blades provide larger energy transfer (higher value of absolute velocity, c_2, at the outlet) as compared to impellers with backward-curved blades. High value of absolute velocity at the outlet is not needed and it is difficult to convert it to pressure. Figure 3.5c shows outlet velocity triangles for impellers with backward-curved blades. The value of whirl component at outlet is low for backward-curved blades. This results in low energy transfer for a given impeller tip speed.

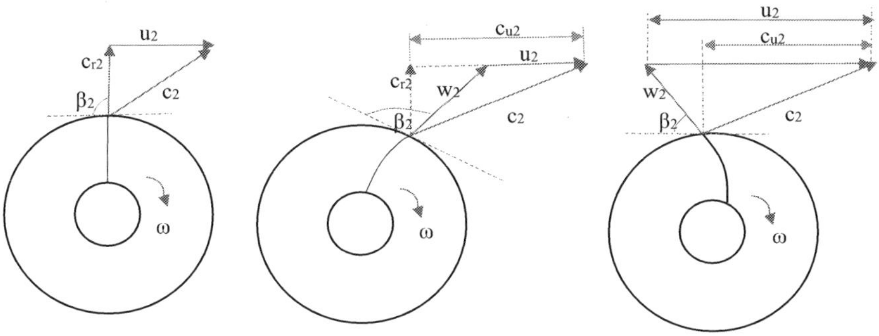

FIGURE 3.5 Outlet velocity triangles for different blade orientation: (a) radial; (b) forward-curved; and (c) backward-curved.

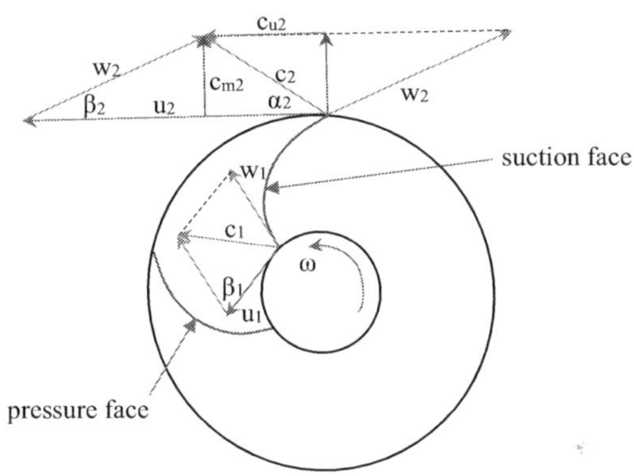

FIGURE 3.6 Velocity triangles at the inlet and outlet of a backward-curved impeller of a centrifugal pump.

3.6 PROCEDURE FOR DRAWING VELOCITY TRIANGLES

Figure 3.6 shows velocity triangles at the inlet and outlet of a backward-curved impeller of a centrifugal pump. The following procedure can be used as a general guideline in drawing velocity triangles:

1. The tangential velocity component, u, is drawn tangent to the impeller at the blade outlet.
2. The relative velocity component, w, is drawn in the direction of the path of the fluid.
3. The absolute velocity, c, is the vector sum of the tangential and the relative velocity, i.e., $c = u + w$.
4. The tangential component of the absolute velocity is the projection of the absolute velocity in the tangential velocity direction.
5. Subscripts 1 and 2 are assigned for inlet and outlet, respectively.

3.7 EULER'S EQUATION

In the following we will develop a one-dimensional equation suitable for the design of a pump. Consider Figure 3.3; from the velocity triangles using cosine law we can write:

$$w_1^2 = c_1^2 + u_1^2 - 2c_1u_1 \cos\alpha_1 \ ; \ c_{u1} = c_1 \cos\alpha_1 \tag{3.15}$$

$$w_2^2 = c_2^2 + u_2^2 - 2c_2u_2 \cos\alpha_2 \ ; \ c_{u2} = c_2 \cos\alpha_2 \tag{3.16}$$

Substituting Equations 3.15 and 3.16 for Equation 3.14 and modifying, we obtain Euler's equation for pumps (Dixon and Hall 2013):

$$Y_{t\infty} = u_2c_{u2} - u_1c_{u1} \tag{3.17}$$

Euler's Equation (3.17) gives the ideal energy transfer to the fluids without considering any losses. It assumes an infinitely thin, infinite number of blades, where the channels between the blades are so narrow that the liquid follows the shape of the blade exactly. This assumes that the fluid is perfectly guided from inlet to outlet. The subscript t∞ indicates the ideal case of congruent streamlines with perfect guidance from inlet to outlet, i.e., an infinitely thin, infinite number of blades. Equation 3.17 can be rewritten in the following form (Equation 3.18) to give theoretical head rise. It has a unit of length (m) or (ft).

$$H = \frac{Y_{t\infty}}{g} = \frac{u_2c_{u2} - u_1c_{u1}}{g} \tag{3.18}$$

3.8 WORKING EQUATION OF A PUMP WITH INFINITELY THIN AND INFINITE NUMBER OF BLADES

Considering Equation 3. 17, and the velocity triangle at the inlet (Figure 3.7), when $c_{u1} = 0$, we get:

$$Y_{t\infty} = u_2c_{u2} \tag{3.19}$$

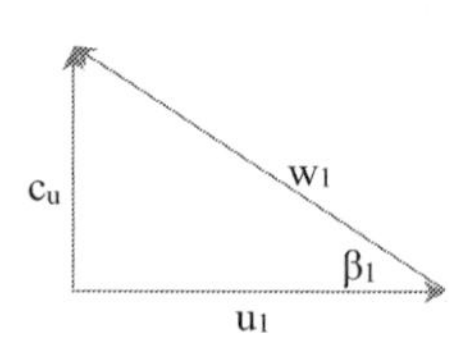

FIGURE 3.7 Velocity triangle at the inlet.

Considering the velocity triangles shown in Figure 3.3, we get:

$$c_{u2} = u_2 - \frac{c_{m2}}{\tan \beta_2} \tag{3.20}$$

Substituting Equation 3.20 for Equation 3.17 and modifying we get:

$$Y_{t\infty} = u_2^2 - \frac{u_2 c_{m2}}{\tan \beta_2} \tag{3.21}$$

Equation 3.21 is the equation of a pump with an infinite number of infinitely thin impeller blades and describes energy increase of an ideal fluid. This energy increase is obtained assuming a uniform velocity distribution along the outlet and the inlet cross-section of the impeller (Figure 3.8). Figure 3.8 shows the relative velocity and pressure along the width of the impeller channel for two cases: (a) for an impeller with an infinite number of blades; and (b) with a finite number of blades. Imagine the impeller as being made with an infinite number of infinitesimally thin blades. Then an ideal flow would be one which is perfectly guided by the blades and would leave the impeller at the blade angle. For the ideal case, streamlines are congruent with the impeller blades, hence the angle the relative velocity w_2 makes with the tangential velocity u_2 is the same as the blade angle, i.e., $\beta_2 = \beta'_2$ (Figure 3.9). For an ideal fluid we assume that all streamlines are congruent, within an infinitely thin, infinite number of blades guiding the flow perfectly in between blades from inlet to the outlet. The head obtained in such a case is what we call theoretical head, given by Equation 3.18.

The meridional component of the absolute velocity c_{m2} is perpendicular to the tangential velocity component. For a finite number of blades, the angle the relative velocity

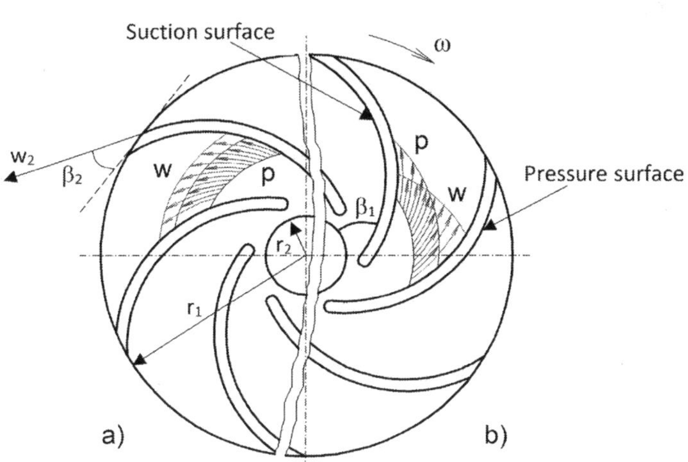

FIGURE 3.8 Diagram of the relative velocity and pressure along the width of the impeller channel.

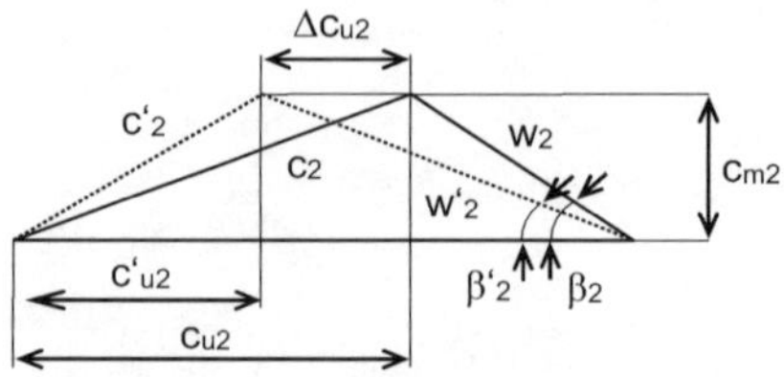

FIGURE 3.9 Velocity triangle at the exit.

w_2 makes with the tangential velocity u_2, $\beta'_2 < \beta_2$ of the blade angle (Figure 3.9). This is due to inability of the blades to perfectly guide the flow. As a result, the C'_{u2} is less than C_{u2}, consequently reducing the theoretical head produced. It should be noted that this decrease of the theoretical head is not a real hydraulic loss.

This change in speed is shown in the exit velocity triangle (Figure 3.9). Then the difference in energy per unit mass between impellers of infinite and finite number of blades is given by:

$$Y_{t\infty} - Y_t = u_2(c_{u2} - c'_{u2}) = u_2 \Delta c_{u2} \tag{3.22}$$

The theoretical energy imparted to the fluid for an impeller with a finite number of blades is then given by:

$$Y_t = u_2^2\left(1 - \frac{\Delta c_{u2}}{u_2}\right) - u_2 \frac{c_{m2}}{\tan\beta_2} = u_2^2 \sigma_Y - u_2 \frac{c_{m2}}{\tan\beta_2} \tag{3.23}$$

$$\text{Where} \quad \sigma_Y = 1 - \frac{\Delta c_{u2}}{u_2} \tag{3.24}$$

is the correction factor that accounts for the finite number of blades.

For a finite number of blades, the relative flow leaving the impeller of a pump will receive imperfect guidance from the blades. In such cases the flow is said to *slip*. A slip factor is used to specify the flow slip effect in the exit of a pump impeller. A slip factor, μ, may be defined as the ratio of the actual tangential flow velocity to the ideal tangential flow velocity:

$$\mu = \frac{c'_{u2}}{c_{u2}} \tag{3.25}$$

Equation 3.24 in terms of the slip factor can be rewritten as:

$$\sigma_Y = 1 - \frac{c_{u2}}{u_2}(1 - \mu) \tag{3.26}$$

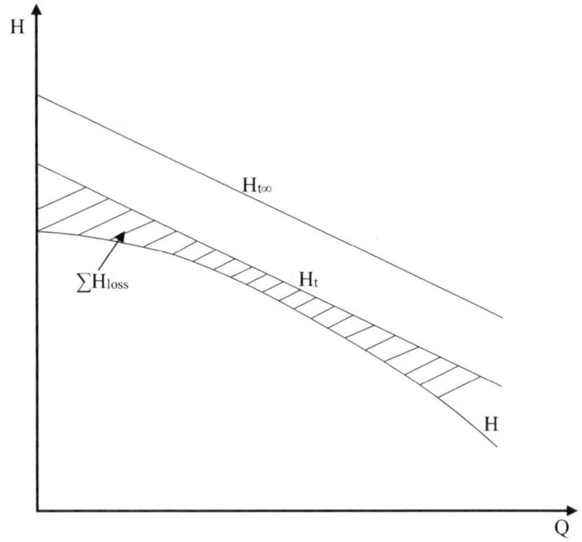

FIGURE 3.10 Theoretical and real characteristics of the pump.

For given impeller and speed the theoretical head H_∞ varies linearly with flow rate Q. The actual flow rate versus head curve is not the same as the theoretical one. This is because for real flow there is fluid friction in the boundary layers along the rotating and fixed passages. Other losses include velocity equalization loss after the blades, and recirculation loss at the inlet. The theoretical head for an impeller with infinitely thin, infinite number of blades, $H_{t\infty}$, the theoretical head for an impeller with finite number of blades, H_∞, and the actual head, H, are shown in Figure 3.10. The shaded region in the figure shows the total losses.

3.8.1 Slip Factor Correlations

Many correlations have been developed to account for a slip factor. In the following we will present some of them.

3.8.1.1 Stodola's Correlation

According to Stodola, nonuniform velocity distribution is caused by the local vortex. This vortex is created in the channel space between the blades (Figure 3.11). The magnitude of the angular velocity of the vortex is equal to the angular velocity of the impeller.

$$\Delta c_{u2} = \frac{a}{2}\omega \tag{3.27}$$

$$\text{Where} \quad a = \frac{2\pi r_2}{z}\sin\beta_2 \tag{3.28}$$

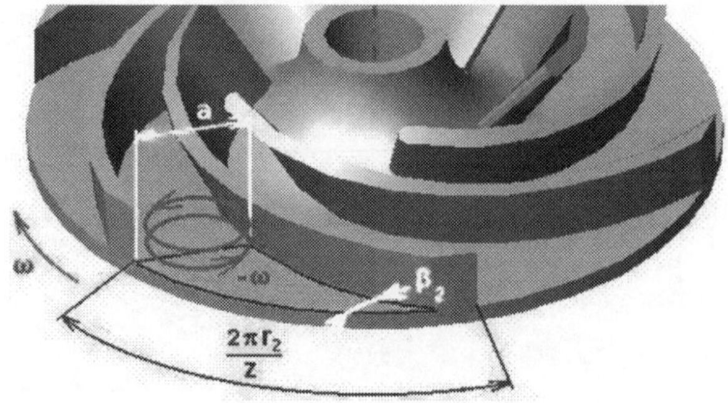

FIGURE 3.11 Vortex in the space between the impeller blades.

then the correction factor is given by

$$\sigma_Y = 1 - \frac{\Delta c_{u2}}{u_2} = 1 - \frac{\pi \sin \beta_2}{z} \tag{3.29}$$

where z is the number of blades. The relationship shows that the slip factor increases with increasing number of blades and decreases with increasing outlet blade angle.

3.8.1.2 Stodola–Serstjuk Correlation (Varchola and Hlobocan, 2016)

Serstjuk modified Stodola's equation as follows:

$$\sigma_Y = 1 - \frac{\Delta c_{u2}}{u_2} = 1 - \frac{\dfrac{2\pi \sin \beta_2}{\sqrt{3}z}}{1 + \dfrac{2\pi \sin \beta_2}{\sqrt{3}z}} \tag{3.30}$$

3.8.1.3 NEL Correlation (Turton, 1984)

NEL correlation takes into consideration vortex at the inlet and outlet of an impeller. The correction factor proposed by NEL is given as follows:

$$\sigma_Y = 1 - \frac{\pi}{z}\left(2 - \frac{D_1}{D_2}\right) - \left(\frac{D_1}{D_2}\right)^2 \sin^2\left(\frac{\beta_1 + \beta_2}{2}\right) \tag{3.31}$$

3.8.1.4 Busemann's Correlation (Buseman, 1928)

Busemann analyzed two-dimensional rotational flow in between impeller blades with the assumption that:

1. infinitely thin blades in the shape of a logarithmic spiral with constant blade angle, β
2. constant width of the meridional section.

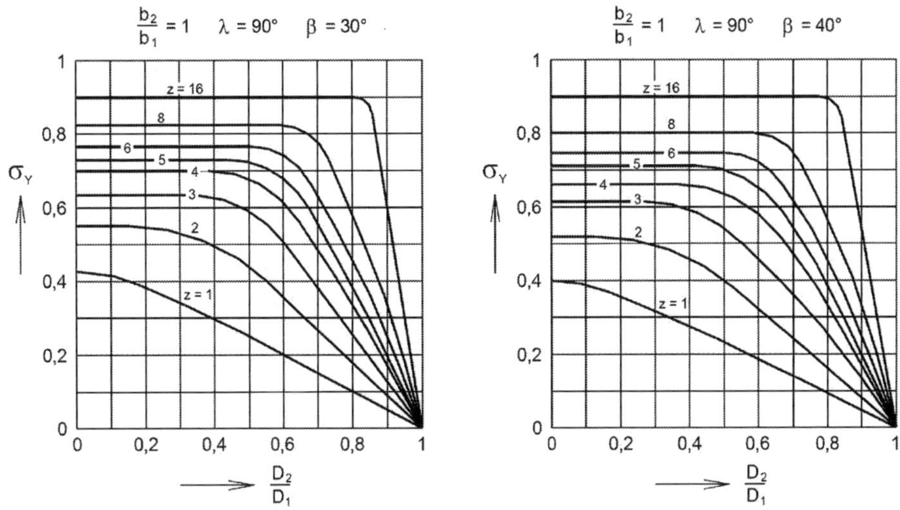

FIGURE 3.12 Correction factors for two blades angles according to Busemann.

Since the resulting equations are complicated for use, the results for two blade angles are shown in graphical form in Figure 3.12.

3.8.1.5 Waisser's Correlation (Rama and Aijaz, 2003)

Waisser described simple analytical expressions for Busemann's graphs.

$$\sigma_Y = a - \frac{b + c\sin\beta_2}{\sqrt{z}} \tag{3.32}$$

where constants a, b, and c are 1.103, 0.523, and 0.582, respectively. Equation 3.32 is valid for a number of blades, $z = 3$–16, and $\beta_2 = 15$–35°.

3.8.2 Fundamental Design Parameters for Hydrodynamic Pumps

Q_n (m^3/s) is the pump rated flow rate, aka capacity
H_n (m) is the rated pump head
P_n (kW) is the rated pump power at the rated flow rate
η_n (%) is efficiency at the rated flow rate
$NPSH_r$ (m) is the minimum pressure required at the suction side of the pump to keep the pump from cavitating
$NPSH_a$ (m) is the available pressure at the suction side of the pump.

3.8.3 Design of Radial Impellers

In the following we will outline the procedure for designing an impeller of radial pumps. The design procedure is applicable to plane-curved impeller blades. Consider Figure 3.13. In the case of a low-speed (small value of specific speed: Figure 3.14) radial impeller, the shape of the meridional section is simple, and the number of main parameters is reduced to a minimum. The low specific speed impeller is designed with plane-curved blades. The main dimensions of the meridional section are shaft diameter d_h, hub diameter D_n, impeller inlet diameter D_0, mean effective impeller diameter at inlet D_{0s}, impeller eye diameter D_1, impeller discharge diameter D_2, inlet edge width b_1, inlet inclination angle γ_1, width of the outlet edge b_2, and the radius of curvature of the rear disc R_d. The angle of inclination of the exit edge γ_2 is usually equal to zero. Also, the angle of inclination of the inlet edges γ_1 is either equal to zero or has values ranging from 13° to 30°. The angle of inclination of the front disc from the vertical, δ, is 0–5°. In some cases, the inclination of the inlet edge γ_1 is equal to zero and width b_1 equal to b_2, which gives a zero angle δ (see Figure 3.13b). In the design of impeller several speed constants are used. These speed constants are designated with K and a subscript. For example, K_{u2} is speed constant, which is a factor describing the relationship between the pump total head and the impeller peripheral velocity, $u_2 = K_u\sqrt{2gH}$. There are several of these constants in use.

For determining the main dimensions of impeller of a pump the following steps are followed.

Shaft diameter, d_h

$$d_h = \sqrt[3]{\frac{5\tau}{\tau_{sp}}} \tag{3.33}$$

where τ is the shaft torque and τ_{sp} is the permissible shaft shear stress.

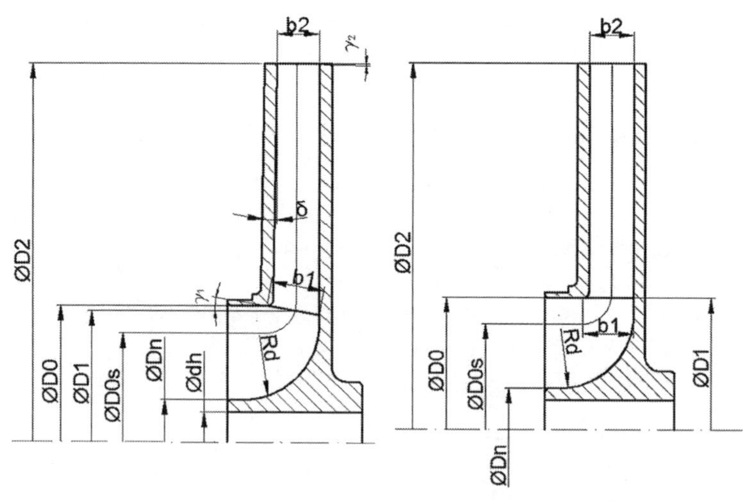

FIGURE 3.13 Main dimensions of low-speed radial pump impeller (a) and (b) simplified version of (a).

Torque, M_k

$$M_k = \frac{\rho Q g H_{total}}{\varpi \eta_c} \tag{3.34}$$

where H_{total} is the total head generated by a multistage pump, η_c is the total efficiency of the pump, ω is the angular speed of the impeller, and ρ is the density of the fluid. The overall efficiency, $\eta_c = \eta_h \eta_m \eta_v$, where η_h is hydraulic efficiency, η_m is mechanical efficiency, and η_v is volumetric efficiency.

For a single-stage pump the torque is given by:

$$M_k = \frac{\rho Q g H}{\varpi \eta_c} \tag{3.35}$$

where H is the head generated by a single-stage pump.

The power input of the pump is calculated from:

$$P = \frac{\rho Q g H_{total}}{\eta_c} \tag{3.36}$$

Hub diameter, D_n

$$D_n = 1.2 d_h \tag{3.37}$$

Coefficient of contraction at the inlet, φ_1

This value is usually selected as 0.75.

Impeller inlet diameter, D_0

$$D_o = \sqrt{\frac{4.16 Q_n}{\pi c_{m1} \varphi_1} + D_n^2} \tag{3.38}$$

Meridional velocity at the inlet, c_{m1}

$$c_{m1} = K_{m1} \sqrt{2 g H_{(1)}} \tag{3.39}$$

where $H_{(1)}$ is the single-stage (one-impeller) head.

If the pump is a multi-stage pump, $H_{(1)}$ is obtained by dividing the total head by the number of impellers, i.e. $\frac{H_{total}}{z}$, where z is the number of impellers.

Coefficient K_{m1} is determined from the following relationships, depending on specific speed n_b. n_b for a single-stage pump is determined from:

$$n_b = n\frac{\sqrt{Q_{(1)}}}{\left(gH_{(1)}\right)^{3/4}} \tag{3.40}$$

where $Q_{(1)}$ is single-stage impeller flow rate, and n is rotational speed in Hz.

For determination of the impeller profile, the meridional velocity at the inlet should be determined as follows:

$$K_{m1} = \frac{c_{m1}}{\sqrt{2gH_{(1)}}} \tag{3.41}$$

$$\begin{aligned} &for\ 0.041 \le n_b \le 0.1647 \\ &K_{m1} = 0.12 + 0.5617\left(n_b - 0.041\right)^{0.8} \\ &for\ n_b > 0.1647 \\ &K_{m1} = 0.227 + 0.574\left(n_b - 0.1647\right) \end{aligned} \tag{3.42}$$

Impeller eye diameter, D_1

$$D_1 = D_{os} + K_1\left(D_o - D_{os}\right) \tag{3.43}$$

where K_1 is the coefficient used to calculate D_1

Mean effective impeller diameter at inlet, D_{0s}

$$D_{os} = \sqrt{\frac{D_0^2 + D_n^2}{2}} \tag{3.44}$$

Coefficient, K_1

$$\begin{aligned} &for\ n_b \le 0.1647 \\ &K_1 = 0.59 + 0.47\cos\left(1225 n_b\right) \\ &for\ n_b > 0.1647 \\ &K_1 = 0 \end{aligned} \tag{3.45}$$

Impeller discharge diameter, D_2

$$D_2 = \frac{2}{\varpi} K_{u2} \sqrt{2gH_{(1)}} \sqrt{\frac{0.84}{\eta_h}} \tag{3.46}$$

where η_h is hydraulic efficiency of the pump and ω is in rad per second. There are several empirical relationships proposed by researchers for determining hydraulic efficiency. A few of them are listed below. According to Wislicenus: $\eta_h = \sqrt{\eta_c} - (0.2 \ to \ 0.6)$, where η_c is the overall pump efficiency. Again, there are several empirical relationships and charts for determining the η_c. For flow rates up to 0.65 m³/s the following relationships can be used to calculate the overall efficiency (Strýček 1994).

$$\begin{aligned}
&for \ 0.04 \le n_b \le 0.33, \quad \eta_c = \sqrt{\left(\frac{Q}{0.048}\right)^{0.083} - \left(-0.722 - \log n_b\right)^3} - \xi \\
&for \ n_b > 0.33, \quad \eta_c = \sqrt{\left(\frac{Q}{0.048}\right)^{0.07} - \left(-0.722 - \log n_b\right)^5} - \xi \\
&\xi = 0.2 \ for \ single \ stage \ pumps \\
&\xi = 0.22 \ for \ multistage \ pumps
\end{aligned} \tag{3.47}$$

For flow rates greater than or equal to 0.65 m³/s, the following relationships can be used to calculate the overall efficiency (Strýček 1994).

$$\begin{aligned}
&for \ 0.04 \le n_b \le 0.33, \quad \eta_c = \sqrt{1.24 - \left(-0.722 - \log n_b\right)^3} - \xi \\
&for \ n_b > 0.33, \quad \eta_c = \sqrt{1.24 - \left(-0.722 - \log n_b\right)^5} - \xi \\
&\xi = 0.2 \ for \ single \ stage \ pumps \\
&\xi = 0.22 \ for \ multistage \ pumps
\end{aligned} \tag{3.48}$$

Another relationship, developed for $Q > 0.178$ m³s⁻¹, is given below.

$$\eta_c = 1 - \frac{0.06 + 0.304 n_b + \dfrac{43.2}{\left(332.6 n_b + 8.2\right)^2} + \dfrac{0.178 - Q}{2.13 + 66Q}}{Q^{\frac{1}{8}}} \tag{3.49}$$

The overall efficiencies can also be obtained from Figures 3.14–3.16. These figures show variation of overall efficiency with specific speed for different flow rates. Figure 3.17 shows different types of pumps for different specific speeds. Pure radial pumps have the lowest specific speeds. Axial pumps have the highest specific speeds.

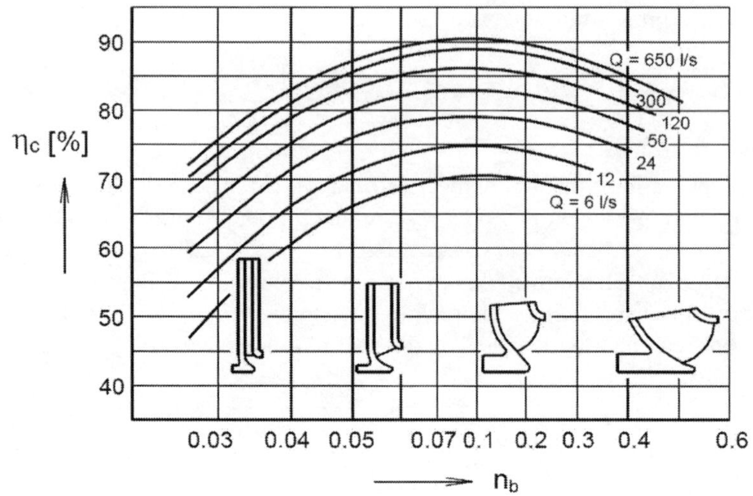

FIGURE 3.14 Variation of pump type and overall efficiency with specific speed.

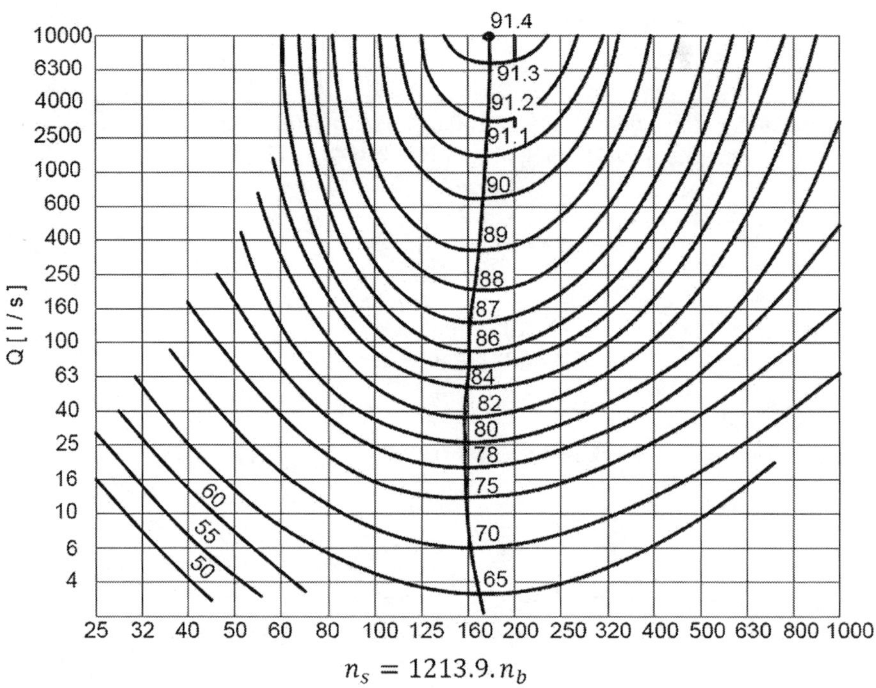

FIGURE 3.15 Overall efficiency of the pump (Krouza 1956).

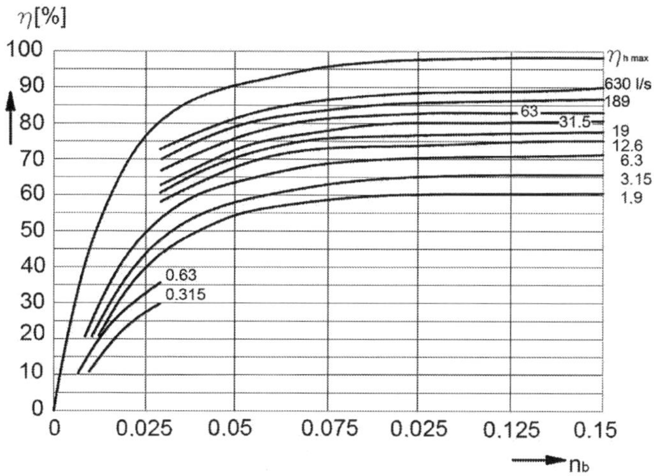

FIGURE 3.16 Overall efficiency of the pump (Karassik 1976).

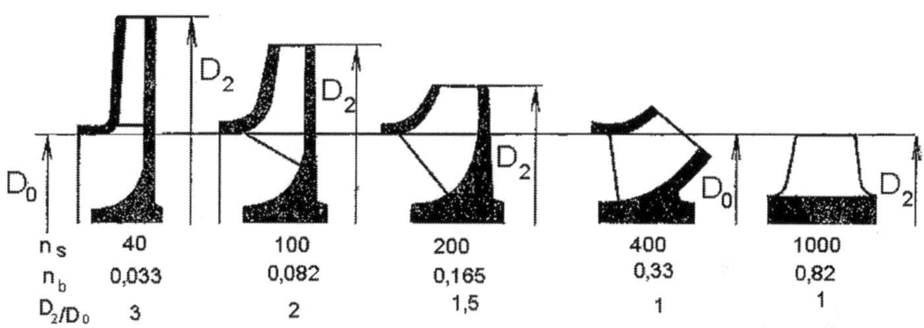

FIGURE 3.17 Types of pumps for different specific speeds.

Other additional empirical relationships for calculating hydraulic efficiency are given below.

$$\eta_h = 1 - \frac{0.42}{\left[\log\left(4200\sqrt[3]{\frac{Q}{60n}}\right) - 0.172\right]^2} \tag{3.50}$$

$$\eta_h = 0.7 + 0.0835 \log\left(4000\sqrt[3]{\frac{Q}{60n}}\right) \tag{3.51}$$

Speed constant K_{u2} describes the relationship between the pump total head and the impeller peripheral velocity. It can be calculated as follows:

$$\begin{aligned} &for\ n_b \leq 0.1 \\ &K_{u2} = (0.875\ to\ 0.89) + 1.1 n_b \\ &for\ n_b > 0.1 \\ &K_{u2} = (0.89\ to\ 0.92) + 3.47 n_b^{1.5} \end{aligned} \tag{3.52}$$

Inlet edge width, b_1

$$b_1 = \frac{1.04 Q_n}{\pi D_1 c_{m1} \varphi_1} \tag{3.53}$$

Note that in this equation 4% increase in flow rate is required to account for volumetric losses.

Coefficient contraction at the outlet, φ_2

This value is usually selected as 0.9.

Meridional velocity at the outlet, c_{m2}

$$c_{m2} = K_{m2} \sqrt{2gH_{(1)}} \tag{3.54}$$

Coefficient K_{m2} is determined from the following relationships:

$$\begin{aligned} &for\ 0.041 \leq n_b \leq 0.1647 \\ &K_{m2} = 0.077 + 1.41 (n_b - 0.041)^{1.2} \\ &for\ n_b > 0.1647 \\ &K_{m2} = 0.192 + 0.8 (n_b - 0.1647)^{1.1} \end{aligned} \tag{3.55}$$

Outlet edge width, b_2

$$b_2 = \frac{1.04 Q_n}{\pi D_2 c_{m2} \varphi_2} \tag{3.56}$$

3.8.4 Design of Blades

Once the main dimension of the impeller is determined, the next step is to determine blade profile (Figure 3.18). Consider the blade profile shown below, where Ψ shows the blade angle distribution. The blade profile should be designed in such a way that boundary layer separation in the flow passage is avoided or delayed. Figure 3.19 shows an impeller with plane-curved blades.

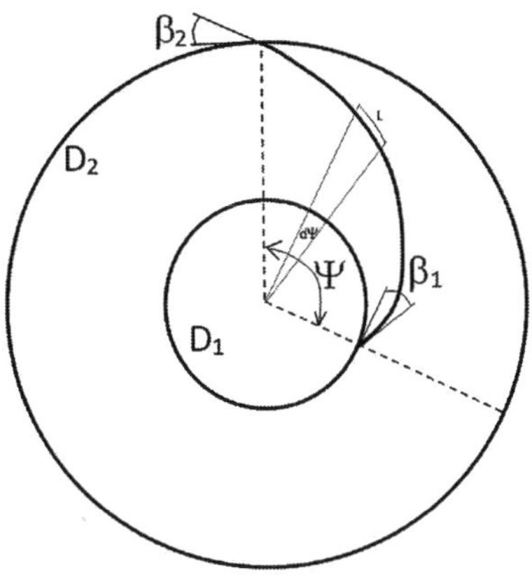

FIGURE 3.18 Creation of blade profile.

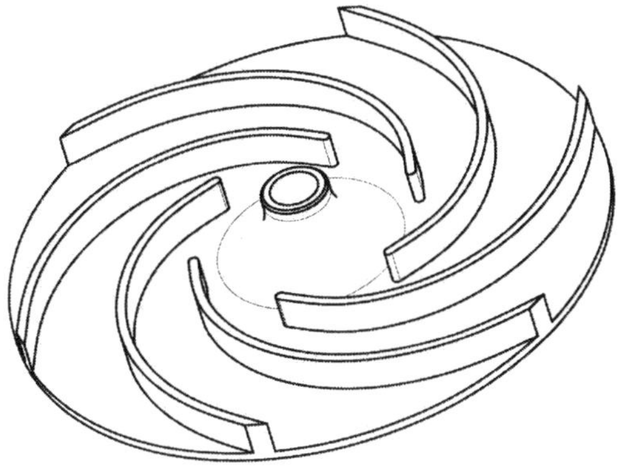

FIGURE 3.19 Example of an impeller with plane-curved blades.

In the following we will describe an iterative approach to determining the blade profile. First, the number of blades and the thickness of the blade are determined using Equations 3.57 and 3.58. The determination of the blade angle β_1 at the inlet is performed iteratively. At the beginning of the iteration, the initial value of the angle β_1 is selected. The value of the contraction factor φ_1 is determined (Equation 3.59), meridional velocity at the inlet c_{m1} is calculated (Equation 3.60), β_1 is calculated (Equation 3.61) and checked if it is the same as the initial estimate. In most cases, agreement is not reached on the first try. Therefore, iteration is necessary until convergence is reached. Once β_1 is determined the next step is to determine β_2 iteratively. We start the iteration by guessing the initial value of β_2. Next, we calculate the contraction factor coefficient φ_2 (Equation 3.62) and meridional velocity c_{m2} (Equation 3.63). Next, we calculate the finite number of blades correction factor for the final number of blades σ_y (Equation 3.64), using the estimated value of β_2. There are several empirical relationships developed for σ_y. Equation 3.64 is one of them. Next, we will determine the head generated by a single-stage pump (Equation 3.65). If the calculated

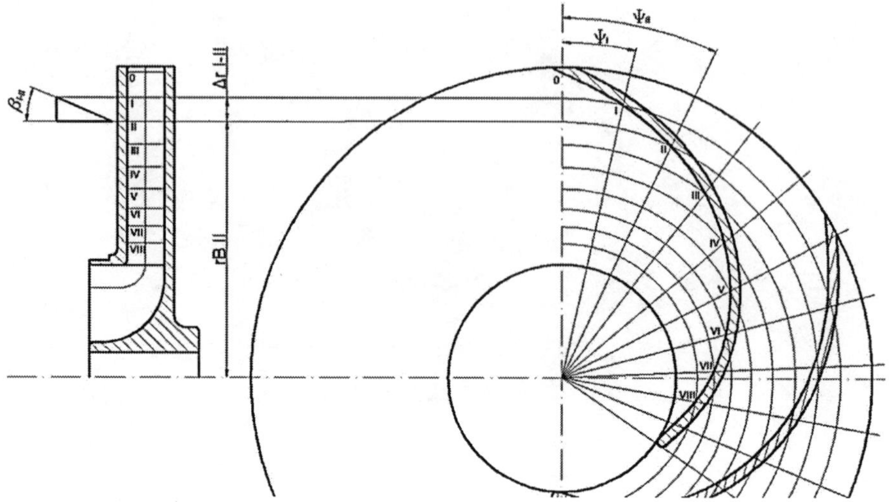

FIGURE 3.20 Principle of formation of blade sections of a slow-running radial impeller.

value of the head does not match the required head value, the initial estimated β_2 value must be changed and the whole cycle is repeated until the required head value is achieved. The equations for determining blade profile are given below (Figure 3.20).

Number of blades, z

$$z \cong 6.5\frac{D_2 + D_1}{D_2 - D_1}\sin\left(\frac{\beta_1 + \beta_2}{2}\right) \tag{3.57}$$

Blade thickness, s

$$s \cong 0.02D_2 \tag{3.58}$$

Inlet contraction coefficient, φ_1

$$\varphi_1 = 1 - \frac{zs}{\pi D_1 \sin\beta_1} \tag{3.59}$$

Inlet meridional velocity, c_{m1}

$$c_{m1} = \frac{1.04Q_n}{\pi D_1 b_1 \varphi_1} \tag{3.60}$$

Inlet blade angle, β_1

$$\beta_1 = \tan^{-1}\left(\frac{c_{m1}}{\pi\frac{D_1}{2}\varpi(1-\kappa)}\right), \quad where \quad \kappa = 0 \ to \ 0.5 \tag{3.61}$$

Outlet contraction coefficient, φ_2

$$\varphi_2 = 1 - \frac{zs}{\pi D_2 \sin\beta_2} \tag{3.62}$$

Outlet meridional velocity, c_{m2}

$$c_{m2} = \frac{1.04Q_n}{\pi D_2 b_2 \varphi_2} \tag{3.63}$$

4% accounts for leakage loss. Correction factor for the final number of blades σ_y

$$\sigma_y = \frac{1 + e^{-\frac{2\pi}{z}\mu\sin\lambda_2} + \left(e^{-\frac{2\pi}{z}\mu\sin\lambda_2}\right)^2}{3}, \quad for\ radial\ impellers\ \lambda_2 = \frac{\pi}{2} \tag{3.64}$$

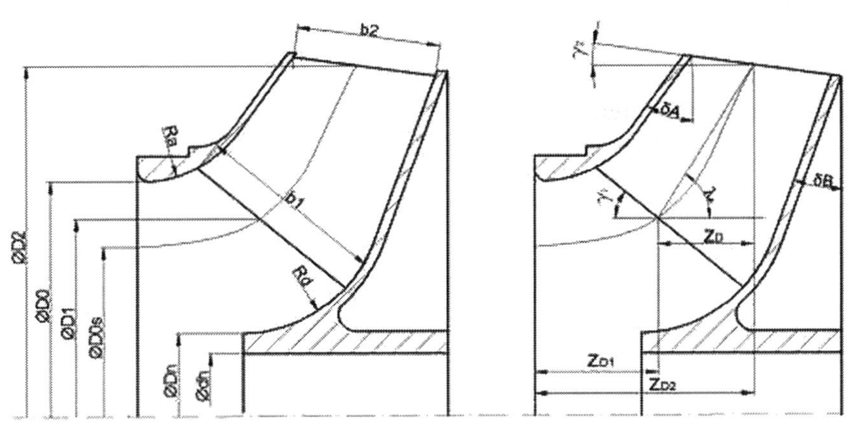

FIGURE 3.21 Main dimensions of a mixed-flow pump. Notice the angle λ.

Head generated by a single stage pump, $H_{(1)}$

$$H_{(1)} = \frac{u_2^2 \sigma_y \eta_h - c_{m2} u_2 \dfrac{1}{\sqrt{\tan\beta_1 \tan\beta_2}} \sqrt[4]{\dfrac{b_2}{b_1}} \sqrt{\dfrac{\varphi_2}{\varphi_1}} \eta_h}{g} \tag{3.65}$$

If the calculated head using Equation 3.65 is not close to the required head, the blade angles are adjusted until a satisfactory result is obtained.

where slip factor, μ

$$\mu = \frac{\tan\beta_2}{1 + \pi \tan\beta_2} \tag{3.66}$$

Blade angle distribution, $\Psi_{(i)}$

$$\Psi_{(i)} = \Psi_{(i-1)} + \frac{\dfrac{\Delta l_{(i)-(i-1)}}{r_{(i-1)} \tan\beta_{(i-1)}} + \dfrac{\Delta l_{(i)-(i-1)}}{r_{(i)} \tan\beta_{(i)}}}{2} \tag{3.67}$$

3.9 MIXED-FLOW PUMPS

3.9.1 Introduction

Mixed-flow pumps are a subcategory of centrifugal pumps, with the fluid entering parallel to the shaft (axis of rotation) and leaving at an angle less than 90° to the shaft, measured from the suction side (Figure 3.22). Mixed-flow pumps are used to bridge the gap between radial and axial flow pumps. In the design of mixed-flow impellers, the equations used in the design of radial impeller can be used to calculate the specific speed, overall efficiency, hydraulic efficiency, torque, and power. Mixed-flow pumps are commonly employed in applications that require a high flow rate with a low discharge pressure.

3.9.2 Working Principles

In mixed-flow pumps, both radial and axial velocity components are present from inlet and outlet. Mixed-flow turbomachines are characterized in that the fluid flows through the impeller at an angle of less than 90° and greater than 0° to the axis of rotation. In general, the mixed-flow impeller can be applied in conjunction with a spiral casing (Figure 3.22a) and for a multi-stage pump with a diffuser (Figure 3.22b). In mixed-flow pumps discharge is at some intermediate angle, between 0° and 90°. Mixed-flow pumps can handle proportionally greater volumes at a medium range of pressures. In mixed-flow pumps, the meridional velocity at the entrance is not the same as the meridional velocity at discharge; usually, $c_{m1} > c_{m2}$.

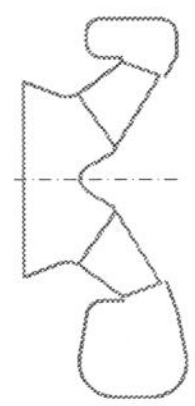

(a) Mixed-flow impeller with spiral casing impeller

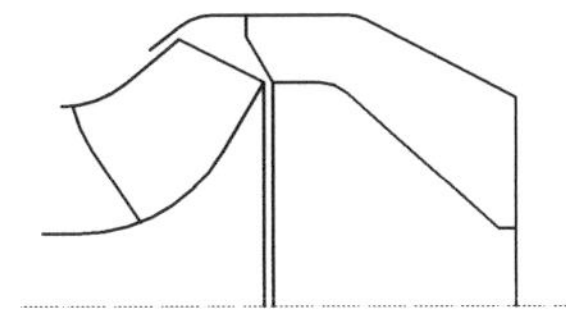

(b) Vane diffuser with mixed-flow

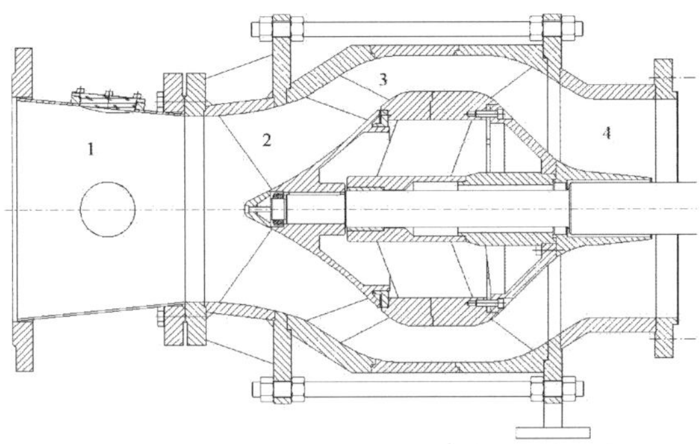

(c) Mixed-flow pump parts: sectional view

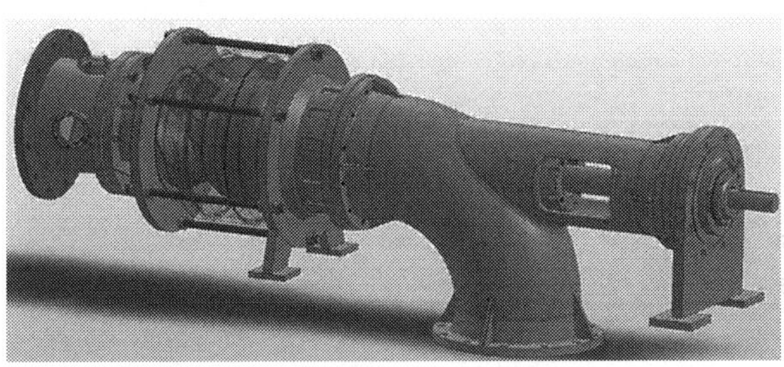

(d) View of the mixed-flow pumpshown in Fig. 3.22a

FIGURE 3.22 Different arrangements of mixed-flow pumps.

Figure 3.22c displays the main parts of the mixed-flow pump with diffuser. The main parts are: (1) inlet cone; (2) mixed-flow impeller; (3) vane diffuser; and (4) outlet area.

3.9.3 Hydraulic Design of Impeller

Similar to radial pumps, the hydraulic design of a mixed-flow impeller is based on the required pressure and flow rate as described in Section 3.8.3. It begins with the determination of the specific speed. The equations described in Section 3.8.3 can be used to calculate the specific speed, overall efficiency, hydraulic efficiency, torque, and power. The main dimensions of the meridional section of a mixed-flow impeller

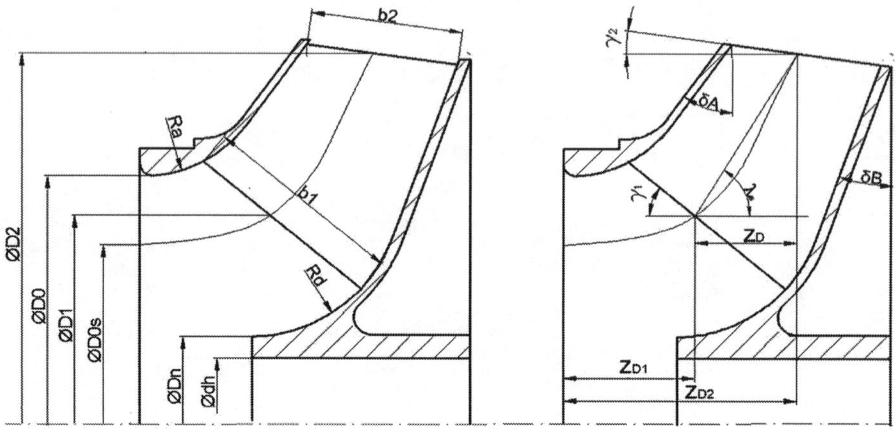

FIGURE 3.23 Main dimensions of the meridional section of a mixed-flow impeller.

are shown in Figure 3.23. These dimensions are: shaft diameter d_h, hub diameter D_n, inlet diameter D_0, diameter at the center streamline D_{0s}, diameter at the intersection of the center streamline and the inlet edge of the blade D_1, diameter at the intersection of the center streamline and the outlet edge D_2, inlet edge width b_1, inlet edge inclination angle γ_1, outlet edge width b_2, the angle of inclination of the outlet edge γ_2, the length of the center streamline *S-S* from the inlet to the outlet (shown as Z_D in the axial direction).

The specific shape of the meridional section depends mainly on the specific speed of the proposed pump and other factors of the design, whether scroll casing is used or if it is a multi-stage pump, and requirements for suction capacity, noise, and vibration. For the hydraulic design of the main dimensions (with the exception of D_1, b_1, b_2, Z_D), we proceed, in principle, in the same way as in the case of the radial impeller (section 3.8.3). When designing the meridional section, we assume uniform meridional velocity distribution along the inlet and outlet edges. We call these velocity values the mean values of the meridional velocity. The angle γ changes linearly from input to output, so its values changes from γ_1 to γ_2. Table 3.1 summarizes the equations used in the design of a mixed-flow impeller.

3.10 EXAMPLE PROBLEMS

Example 3.1

A centrifugal pump's tip speed is 10 m/s while the flow velocity is 1.2 m/s, discharging water at a flow rate of 60 l/s. The blade angle at the outlet is 25° to the tangent at the impeller periphery. If the impeller radius is 0.5 m, for fluid entering in the axial direction with no slip, calculate the torque delivered by the impeller.

TABLE 3.1
Calculation of the Main Dimensions of the Meridional Section of a Diagonal Impeller

Quantity	Symbol	Equation	Equation Number
Torque	M_k	$M_k = \dfrac{P}{2\pi n_n}$	(3.65)
Shaft diameter	d_h	$d_h = \sqrt[3]{\dfrac{5M_k}{\tau_{dov}}}$	(3.66)
Hub diameter	D_n	$D_n = 1.2d_h$	(3.67)
Contraction coefficient at the inlet	ϕ_1	Usually taken as 0.75	
Constant	K_{m1}	For $n_b > 0.1647$ $K_{m1} = 0.27 + 0.574(n_b - 0.1647)$	(3.68)
Mean meridional velocity	c_{m1}	$c_{m1} = K_{m1}\sqrt{2(gH)_{(1)}}$	(3.69)
Volumetric flow rate loss	$\sum Q_z$	$0.04(Q_n)$	(3.70)
Inlet impeller diameter	D_0	$D_o = \sqrt{\dfrac{4(Q_n + \sum Q_z)}{\pi c_{m1}\phi_1} + D_n^2}$	(3.71)
Diameter at the intersection of the center streamline and the inlet edge of the blade	D_1	$D_1 = 2\sqrt{\dfrac{D_0^2}{4} - \dfrac{(Q_n + \sum Q_z)\sin\gamma_1}{2\pi c_{m1}\phi_1}}$	(3.72)
Diameter at the center streamline	D_{0s}	$D_{os} = \sqrt{\dfrac{D_o^2 + D_n^2}{2}}$	(3.73)
Constant	K_{u2}	For $n_b > 0.1$: $K_{u2} = (0.89 - 0.92) + 3.47n_b^{1.5}$	(3.74)
Diameter at the intersection of the center streamline and the outlet edge	D_2	$D_2 = \dfrac{2}{\pi n_n} K_{u2}\sqrt{2(gH)_{(1)}}\sqrt{\dfrac{0.84}{\eta_h}}$	(3.75)
Radius at the inlet blade tip	R_{A1}	$R_{A1} = \sqrt{\dfrac{D_1^2}{4} + \dfrac{(Q_n + \sum Q_z)\sin\gamma_1}{2\pi c_{m1}\phi_1}}$	(3.76)

TABLE 3.1 (Continued)
Calculation of the Main Dimensions of the Meridional Section of a Diagonal Impeller

Quantity	Symbol	Equation	Equation Number
Radius at the inlet blade hub	R_{B1}	$R_{B1} = \sqrt{\frac{D_1^2}{4} - \frac{(Q_n + \sum Q_z)\sin\gamma_1}{2\pi c_{m1}\phi_1}}$	(3.77)
Inlet edge width	b_1	$b_1 = \frac{R_{A1} - R_{B1}}{\sin\gamma_1}$	(3.78)
Contraction coefficient at the outlet	ϕ_2	take 0.9	
Meridional velocity at the outlet	c_{m2}	$c_{m2} = K_{m2}\sqrt{2(gH)_{(1)}}$	(3.79)
Coefficient	K_{m2}	for $n_b > 0.1647$: $K_{m2} = 0.192 + 0.8(n_b - 0.1647)^{1.1}$	(3.80)
Outlet edge width	b_2	$b_2 = \frac{(Q_n + \sum Q_z)}{\pi D_2 c_{m2}\phi_2}$	(3.81)

Solution

Given: $u_2 = 10$ m/s, $c_{m2} = 1.5$ m/s, zero slip condition $b_2 = b'_2 = 28°$

From the theoretical pump head equation. Since $c_{u1} = 0$

$$H = \frac{Y_{t\infty}}{g} = \frac{u_2 c_{u2} - u_1 c_{u1}}{g} = \frac{u_2 c_{u2}}{g}$$

From the outlet velocity triangle (Figure 3.24),

$$\text{we get: } \tan\beta_2 = \frac{c_{m2}}{u_2 - c_{u2}} \Rightarrow u_2 - c_{u2} = \frac{c_{m2}}{\tan\beta_2} \Rightarrow c_{u2} = u_2 - \frac{c_{m2}}{\tan\beta_2}$$

Therefore:

$$H = \frac{Y_{t\infty}}{g} = \frac{u_2 c_{u2}}{g} = \frac{u_2}{g}\left(u_2 - \frac{c_{m2}}{\tan\beta_2}\right) = \frac{10\frac{m}{s}}{9.806\frac{m}{s^2}}\left(10\frac{m}{s} - \frac{1.2\frac{m}{s}}{\tan(25^o)}\right) = 7.573m$$

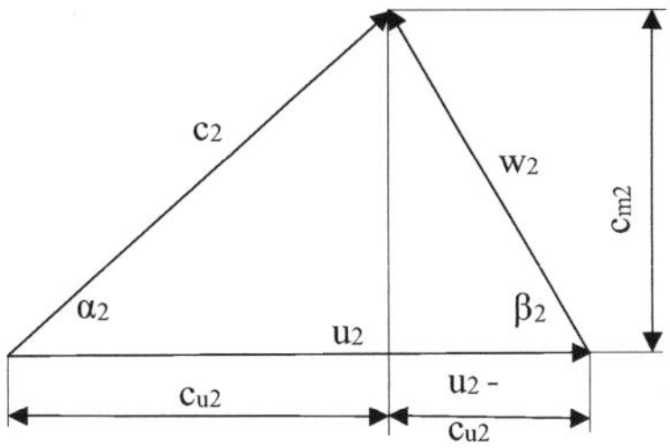

FIGURE 3.24 Outlet velocity triangles.

$$\text{Power: } P = \rho QgH = 998\frac{kg}{m^3}\cdot 0.06\frac{m^3}{s}\cdot 9.806\frac{m}{s^2}\cdot 7.573m = 4447W$$

$$\text{Torque: } \tau = \frac{P}{\varpi} = \frac{P}{\frac{u_2}{r}} = \frac{P\cdot r}{u_2} = \frac{4447W\cdot 0.5m}{10\frac{m}{s}} = 222.352N\cdot m$$

Example 3.2

Determine the overall and hydraulic efficiency for a centrifugal pump having a capacity of 2 l/s, at a head of 30 m and rotational speed of 750 rpm. Determine also the type of pump using Figure 3.14.

Given

$$Q_n = 0.002\frac{m^3}{s}, \qquad n_n = 750rpm \qquad H_n = 30m$$

Specific speed

$$n_b = n_n\frac{\sqrt{Q_n}}{(gH_n)^{\frac{3}{4}}} = 0.049$$

$$For\, Q < 0.65\frac{m^3}{s}, \quad 0.04 \le n_b \le 0.33$$

Overall efficiency

$$\xi = 0.2\text{ - }0.3 \quad take\ 0.2\ for\ single\ stage\ pump$$
$$For \ \ 0.04 \le n_b \le 0.33$$

$$\eta_c = \sqrt{\left(\frac{Q_n}{0.048}\right)^{0.083} - \left(\left|-0.722 - \log(n_b)\right|\right)^3} - \xi = 0.697 = 69.7\%$$

Hydraulic efficiency

$$U\sin g\ Wislicenus\ equation$$

$$\eta_h = \sqrt{\eta_c} - 0.04 = 0.795 = 79.5\%$$

Pump type: radial.

EXAMPLE 3.3

The impeller of a pump shown in Figure 3.25 rotates at 1200 rpm and discharges 200 l/s. Determine the radial velocities at the inlet and outlet assuming no whirl at the inlet.

Given

$$N = 1200\,rpm,\ \ Q = 200\frac{l}{s},\ \ D_1 = 100mm,\ \ D_2 = 250mm\ \ \ b_1 = b_2 = 20mm$$

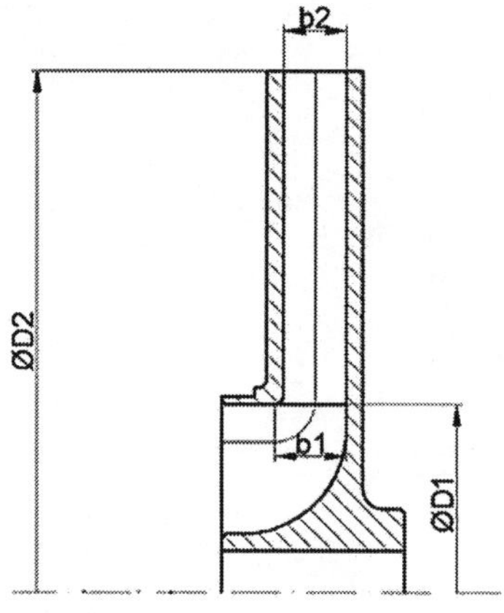

FIGURE 3.25 Outlet velocity triangles.

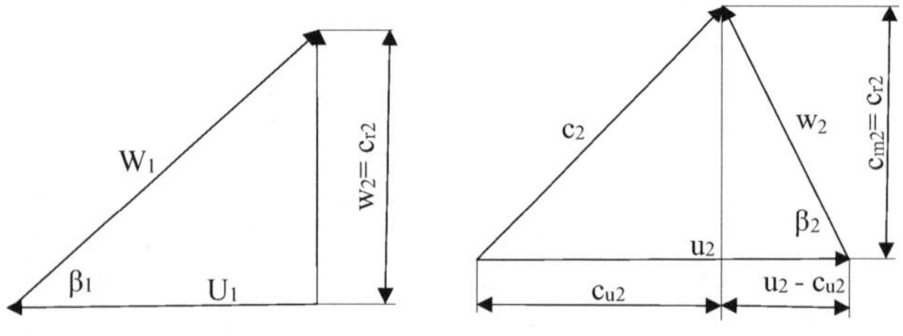

FIGURE 3.26 Inlet and outlet velocity triangles.

Solution

$$\varpi = 125.664 s^{-1}$$

$$u_1 = \varpi \frac{D_1}{2} = 6.283 \frac{m}{s}$$

$$u_2 = \varpi \frac{D_2}{2} = 15.708 \frac{m}{s}$$

$$A_1 = \pi b_1 D_1 = 0.006 m^2$$

$$A_2 = \pi b_1 D_2 = 0.016 m^2$$

$$C_{r1} = \frac{Q}{A_1} = 31.831 \frac{m}{s}$$

$$C_{r2} = \frac{Q}{A_2} = 12.732 \frac{m}{s}$$

Example 3.4

In the following we will show, an example of the hydraulic design of a radial impeller with plane-curved blades for the following nominal parameters:

$Q_n = 50$ l/s
$gH = 500$ J/kg
$n_n = 1450$ rpm

Given:

$$Q_n = 50 \frac{l}{s} \quad Y_n = 500 \frac{J}{kg} \quad H_n = 50.986 m \quad n_n = 24.16 Hz$$

Specifc speed

$$n_b = n_n \frac{\sqrt{Q_n}}{(gH_n)^{\frac{3}{4}}} = 0.051 \quad low-speed\ radial\ impeller,\ see\ Fig.\ 3.14$$

$$n_s = 1213 n_b = 61.975 \quad see\ Fig.\ 3.15$$

$$For\ Q < 0.65 \frac{m^3}{s} \quad 0.04 \le n_b \le 0.33$$

Overall efficiency

$$\xi = 0.2 - 0.3 \quad take\ 0.2\ for\ single\ stage\ pump$$
$$For\ \ 0.04 \le n_b \le 0.33$$
$$\eta_c = \sqrt{\left(\frac{Q_n}{0.048}\right)^{0.083} - \left(\left|-0.722 - \log(n_b)\right|\right)^3} - \xi = 0.704$$

Note that the overall efficiency can also be estimated from Fig. 3.14

Hydraulic efficiency

$$Use\ Wislicenus\ equation$$
$$\eta_h = \sqrt{\eta_c} - 0.04 = 0.799$$

Torque

$$\tau = \frac{\rho Q_n g H_n}{\varpi \eta_c} = 233.882 Nm,\ where\ \varpi = 151.767 s^{-1},\ \rho = 1000 \frac{kg}{m^3}$$

Power

$$P = \frac{\rho Q_n g H_n}{\eta_c} = 3.55\, x 10^4\, W$$

Shaft diameter

$$d_h = \sqrt[3]{\frac{5\tau}{\tau_{allowable}}} = 0.043m,\ \ where\ \ \tau_{allowable} = 15MPa.\ Take\ 45mm$$

Hub diameter

$$D_n = 1.2 d_h = 0.051m \qquad Take\ 55mm$$
$$for\ \ 0.041 \le n_b \le 0.1647$$
$$K_{m1} = 0.12 + 0.5617\left(n_b - 0.041\right)^{0.8} = 0.134$$

Inlet meridional velocity

$$c_{m1} = K_{m1}\sqrt{2gH_n} = 4.241 \frac{m}{s}$$

$$Initially\ select\ the\ inlet\ contraction\ coeficient,\ \varphi_1 = 0.75$$

Impeller inlet diameter

$$D_o = \sqrt{\frac{(4)(1.04)(Q_n)}{\pi(c_{m1})\varphi_1}} = 0.154m$$

We will take 140 mm. This is because $D_1 = D_0$. See Figure 3.27 and steps below.

$$Coefficient\ K_1\ for\ n_b < 0.1647$$
$$K_1 = 0.59 + (0.47)(\cos(1255 \cdot n_b)) = 0.796$$

Impeller diameter at inlet

$$D_{os} = \sqrt{\frac{D_o^2 + D_n^2}{2}} = \sqrt{\frac{(0.14m)^2 + (0.055)m^2}{2}} = 0.106m$$
$$D_1 = D_{os} + K_1(D_o - D_{os}) = 0.145m$$

Inlet edge width Note that $D_1 = D_o$

$$b_1 = \frac{(1.04) \cdot Q_n}{\pi \cdot D_1 \cdot c_{m1} \cdot \varphi_1} = 0.037m$$
$$K_{u2}\ \ for\ \ n_b \le 0.1$$
$$K_{u2} = (0.875 - 0.89) + 1.1n_b,\ take\, 0.88$$
$$K_{u2} = 0.88 + 1.1n_b = 0.936$$

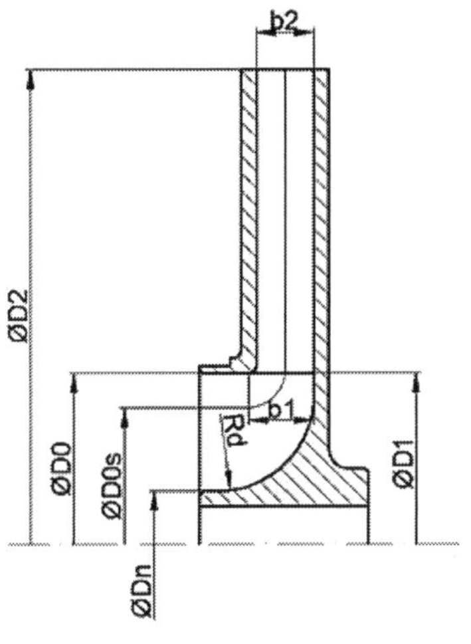

FIGURE 3.27 Cross-sectional view of an impeller showing $D_1 = D_{0.}$

Outlet diameter

$$D_2 = \frac{2}{\varpi} \cdot K_{u2} \cdot \sqrt{2 \cdot g \cdot H_n} \sqrt{\frac{0.84}{\eta_h}} = 0.4m = 400mm$$

$$\textit{Coefficient } K_{m2} \textit{ for } 0.041 \le n_b \le 0.1647$$

$$K_{m2} = 0.077 + 1.41\left(n_b - 0.041\right)^{1.2} = 0.083$$

Meridional velocity at the outlet

$$c_{m2} = K_{m2}\sqrt{2 \cdot g \cdot H_n} = 2.612\frac{m}{s}$$

Coefficient of contraction $\varphi_2 = 0.8$
Initially estimate φ_2 *is chosen between* 0.8 *and* 0.95

$$b_2 = \frac{(1.04) \cdot Q_n}{\pi \cdot D_2 \cdot c_{m2} \cdot \varphi_2} = 0.02m$$

Determination of blade profile

Select blade angle at the inlet $\beta_1 = 26\deg \quad \beta_2 = 21.5\deg$
Blade thickness is usually taken as $s = (0.015 \, to \, 0.025) \cdot D_2$

$$s = 0.02 \cdot D_2 = 0.008m$$

Number of blades

$$z = 6.5 \cdot \frac{D_2 + D_1}{D_2 - D_1} \cdot \sin\left(\frac{\beta_1 + \beta_2}{2}\right) = 5.911 \qquad take\, z = 6$$

Coefficient of contraction at the inlet

$$\varphi_1 = 1 - \frac{z \cdot s}{\pi \cdot D_1 \cdot \sin(\beta_1)} = 0.751, \textit{ note that } z = 6 \textit{ and } D_1 = 140mm \textit{ is taken here}$$

Meridional velocity at the inlet

$$c_{m1} = \frac{1.04 \cdot Q_n}{\pi \cdot D_1 \cdot b_1 \cdot \varphi_1} = 4.919\frac{m}{s}, \textit{with } D_1 = 140mm \textit{ and } b_1 = 32mm$$

Inlet blade angle

$$\beta_1 = \tan^{-1}\left(\frac{c_{m1}}{\frac{D_1}{2}\cdot\omega\cdot(1-\kappa)}\right) = 25.984\deg,\ \ with\ \ \kappa = 0.05$$

$$\varphi_2 = 1 - \frac{z\cdot s}{\pi\cdot D_2\cdot\sin(\beta_2)} = 0.897$$

$$c_{m2} = \frac{1.04\cdot Q_n}{\pi\cdot D_2\cdot b_2\cdot\varphi_2} 2.329\frac{m}{s}$$

$$\mu = \frac{\tan(\beta_2)}{1+\pi\cdot\tan(\beta_2)} = 0.176$$

$$\sigma_y = \frac{1 + e^{\frac{-2\pi}{z}\mu\cdot\sin(\lambda_2)} + e^{\left(\frac{-2\pi}{z}\mu\cdot\sin(\lambda_2)\right)^2}}{3} = 0.932,\ where\ \lambda_2 = \frac{\pi}{2} = 1.571$$

$$H = \frac{u_2^2\cdot\sigma_y\eta_h - u_2\cdot c_{m2}\cdot\frac{1}{\sqrt{\tan(\beta_1)\cdot\tan(\beta_2)}}\cdot\sqrt[4]{\frac{b_2}{b_1}}\sqrt{\frac{\varphi_2}{\varphi_1}}\eta_h}{g} = 56.195m$$

The head is greater than the orginal design value (50m), we can adjust the blade angles to get the desired design value. With $\beta_1 = 15\deg$ *and* $\beta_2 = 22.5\deg$

$$H = \frac{u_2^2\cdot\sigma_y\eta_h - u_2\cdot c_{m2}\cdot\frac{1}{\sqrt{\tan(\beta_1)\cdot\tan(\beta_2)}}\cdot\sqrt[4]{\frac{b_2}{b_1}}\sqrt{\frac{\varphi_2}{\varphi_1}}\eta_h}{g} = 51.866m$$

Example 3.5

Determine the main dimensions of a mixed-flow impeller for a pump with nominal parameters:

$Q_n = 315$ l/s
$gH_n = 225$ J/kg
$n_n = 1850$ rpm

Solution

Given

$$Q_n = 0.315\frac{m^3}{s}\quad Y_n = 225\frac{J}{kg}\quad H_n = \frac{Y_n}{g} = 22.944\quad n_n = 30.83Hz$$

Specific speed

$$n_b = n_n \frac{\sqrt{Q_n}}{\left(g \cdot H_n\right)^{\frac{3}{4}}} = 0.298 \quad \textit{See Fig. 3.11}$$

$$\textit{For } Q < 0.65 \frac{m^3}{s} \qquad 0.04 \le n_b \le 0.33$$

$$\xi = 0.2 - 0.3, \ \textit{take } 0.2 \textit{ for single stage pump}$$

$$\eta_c = \sqrt{\left(\frac{Q_n}{0.048}\right)^{0.083} - \left(\left|-0.722 - \log\left(n_b\right)\right|\right)^3 - \xi} = 0.878$$

Hydraulic efficiency

$$\eta_h = \sqrt{\eta_c} - 0.04 = 0.897$$

Torque

$$M_k = \frac{\rho Q_n g H_n}{\varpi \eta_c} = 416.834 Nm, \ \textit{where } \varpi = 193.72 s^{-1}$$

Power

$$\tau = \frac{\rho Q_n g H_n}{\eta_c} = 80750 W$$

Shaft diameter

$$d_h = \sqrt[3]{\frac{5 \cdot \tau}{\tau_{allowable}}} = 0.052 m \qquad \textit{Take } 45 mm$$

Hub diameter

$$D_n = 1.2 d_h = 0.062 m \quad \textit{Take } 60 mm$$

$$\textit{for } n_b \le 0.1647$$

$$K_{m1} = 0.227 + 0.574\left(n_b - 0.1647\right) = 0.303$$

Inlet meridional velocity

$$c_{m1} = K_{m1}\sqrt{2 g H_n} = 6.437 \frac{m}{s}$$

Contraction coefficient, initially we take $\varphi_1 = 0.8$

Impeller inlet diameter

$$Q_{losses} = 0.04Q_n = 0.013\frac{m^3}{s}$$

$$D_o = \sqrt{\frac{4\cdot(Q_n + Q_{losses})}{\pi\cdot c_{m1}\cdot\varphi_1} + (D_n)^2} = 0.285m$$

Impeller diameter at inlet

$$D_{os} = \sqrt{\frac{D_o^2 + D_n^2}{2}} = 0.206m$$

$$D_1 = 2\sqrt{\frac{D_o^2}{4} - \frac{(Q_n + Q_{losses})\cdot\sin(\gamma_1)}{2\cdot\pi\cdot c_{m1}\cdot\varphi_1}} = 0.219m \quad \gamma_1 = 55\deg$$

$$R_{A1} = \sqrt{\frac{D_1^2}{4} + \frac{(Q_n + Q_{losses})\cdot\sin(\gamma_1)}{2\cdot\pi\cdot c_{m1}\cdot\varphi_1}} = 0.143m$$

$$R_{B1} = \sqrt{\frac{D_1^2}{4} - \frac{(Q_n + Q_{losses})\cdot\sin(\gamma_1)}{2\cdot\pi\cdot c_{m1}\cdot\varphi_1}} = 0.061m$$

Inelt edge width

$$b_1 = \frac{R_{A1} - R_{B1}}{\sin(\gamma_1)} = 0.099m$$

K_{u2} *for* $n_b > 0.1$

$K_{u2} = 0.89 + 3.47n_b^{1.5} = 1.454,$ *note that* $K_{u2} = (0.89 \text{ to } 0.92) + 3.47n_b^{1.5}$

Outlet diameter

$$D_2 = \frac{2}{2\cdot\pi\cdot n_n}K_{u2}\sqrt{2gH_n}\sqrt{\frac{0.84}{\eta_h}} = 0.308m$$

Coefficient K_{m2} *for* $n_b > 0.1647$

$$K_{m2} = 0.192 + 0.8\cdot(n_b - 0.1647)^{1.1} = 0.279$$

Meridional veocity at the outlet

$$c_{m2} = K_{m2}\sqrt{2gH_n} = 5.92\frac{m}{s}$$

Coefficient of contraction, $\varphi_2 = 0.85$

$$b_2 = \frac{(Q_n + Q_{losses})}{\pi\cdot D_2\cdot c_{m2}\cdot\varphi_2} = 0.067m$$

Example 3.6

For a centrifugal pump with an inlet diameter of 70 mm, outlet diameter of 200 mm, inlet blade angle of 30°, and outlet blade angle of 32°, assuming $\alpha_1 = 90°$ determine the head, power, and rotational speed. Assume inlet blade width 5 cm and outlet blade width 3 cm.

$$D_1 := 70\,mm \qquad D_2 := 200\,mm \qquad b_1 := 5\,cm$$

$$D_1 := 0.07\,m \qquad D_2 := 0.2\,m \qquad b_2 := 3\,cm$$

$$\beta_1 := 30\,deg \qquad \beta_2 := 32\,deg \qquad Q := 50\frac{l}{s}$$

$$C_{m1} := \frac{Q}{\pi \cdot D_1 \cdot b_1} = 4.547\frac{m}{s} \qquad C_1 := C_{m1}$$

$$C_{m2} := \frac{Q}{\pi \cdot D_1 \cdot b_1} = 8.703\frac{ft}{s} \qquad C_1 = 4.547\frac{m}{s}$$

$$U_1 := \frac{C_1}{\tan(\beta_1)} = 7.876\frac{m}{s}$$

$$\omega := \frac{U_1}{\frac{D_1}{2}} = 225.032\frac{1}{s}$$

$$U_2 := \omega \cdot \frac{D_2}{2} = 22.503\frac{m}{s}$$

$$W_2 := \frac{C_{m2}}{\sin(\beta_1)} = 5.305\frac{m}{s} \qquad C_2 := \sqrt{U_2^{\,2} + W_2^{\,2} - 2 \cdot U_2 \cdot W_2 \cdot \cos(\beta_2)} = 18.222\frac{m}{s}$$

$$C_{u2} := U_2 - W_2 \cdot \cos(\beta_2) = 18.004\frac{m}{s}$$

$$Y_{t\infty} := U_2 \cdot C_{u2} - U_1 \cdot C_{u1} = 405.152\frac{m^2}{s^2}$$

$$H := \frac{Y_{t\infty}}{g} = 41.314\,m$$

$$P := \rho \cdot Q \cdot g \cdot H = (2.026.10^4)\,W$$

3.11 EXERCISE PROBLEMS

Problem 3.1

The impeller of a pump shown in Figure 3.28 rotates at 1200 rpm. Determine the radial velocities at the inlet and outlet assuming no whirl at the inlet. The pump's outer diameter is 30 cm. The eye diameter is 12 cm. The blade angle at inlet and the

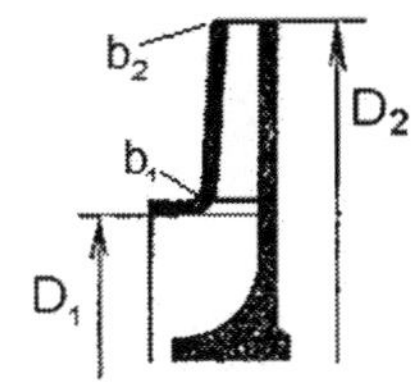

FIGURE 3.28 Impeller cross-sectional view.

outlet are 20° and 25°, respectively. Speed is 1450 rpm. Determine the theoretical head. Also, determine the power if the overall efficiency $\eta_c = 76\%$.

Problem 3.2

A radial centrifugal pump with inlet and outlet diameter of 20 and 60 cm, respectively, delivers water at 45 l/s. Determine the theoretical head generated by the pump. Determine the velocity components and sketch the inlet and outlet velocity triangles.

Problem 3.3

A pump runs at 1400 rpm, and blade angle at the outlet is 26°. The impeller outlet and inlet diameters are 60 and 30 cm, respectively. Water flows at a constant rate of 3 m/s from the inlet to the outlet. Sketch the velocity triangles at the inlet and outlet. Determine the theoretical head, the angle between the absolute velocity of water at outlet and tangential velocity, α_2, the inlet blade angle, β_1, and the specific energy. Also, determine all the velocity components.

Problem 3.4

A centrifugal pump develops a head of 50 m while discharging water at a rate of 0.01 m^3/s. The impeller's diameters at the inlet and outlet are 0.3 and 0.45 m, respectively. The blade angle at the outlet is 28°. The flow area is constant from inlet to the outlet and is equal to 0.05 m^2. Determine the manometric efficiency if the pump runs at 1200 rpm.

Problem 3.5

A single-stage, radial flow centrifugal pump discharges 0.055 m^3/s at a head of 30 m, while running at 1650 rpm. The inlet diameter is120 mm; the outlet diameter 300 mm. The inlet width is 25 mm and the outlet width is 20 mm. The leakage flow is estimated at 1 l/s; mechanical losses are estimated at 1500 W. The contraction factor due to blade thickness is 0.85. The absolute velocity angle at the inlet is 90° and relative velocity angle at the outlet measured from tangential direction 32°. If the overall efficiency of the pump is 60%, determine: inlet blade angle, the angle between the tangential velocity component and the absolute velocity at the outlet, shaft power, and mechanical power. Draw the velocity triangles at the inlet and outlet.

Problem 3.6

A pump discharges 140 l/s at a head of 25 m while rotating at 1550 rpm.

(a) Determine its specific speed.
(b) Determine the type of pump.
(c) If the speed of the pump is changed to 1200 rpm, determine the new head and flow rate.

Problem 3.7

A centrifugal pump has the following specifications:

Outlet diameter =5 cm
Inlet diameter =5 cm
Inlet width = 6 cm
Outlet width = 3 cm
Blade angle at inlet (β_1) = 18°
Blade angle at outlet (β_2) = 15°
N =1400 rpm

Determine: (a) discharge; and (b) head; (c) draw velocity triangles.

Problem 3.8

A centrifugal pump discharges 350 l/s at a head of 40 m. The inlet and outlet blade angles are equal, i.e., $\beta_1 = \beta_2$. Inlet and outlet blade widths are 10 and 5 cm, respectively. Inlet and outlet diameters are 70 and 20 cm, respectively. Assuming $\alpha_1 = 90°$, determine the blade angles and pump speed.

3.12 BIBLIOGRAPHY

Blaha, J. and Brada, K. *Pumping Technology Manual*. Prague: ČVUT, 1997.

Bohl, W. *Fluid Flow Machines, Vol. 2, Calculation and Construction (in German)*. Munich: Vogel, 1999.

Brada, K. and Bláha, J. *Hydraulic Machines (in Czech)*. Prague: SNTL, 1992.

Busemann, Á. "The delivery head ratio of radial centrifugal pumps with logarithmic spiral blades" (in German)," *J. Appl. Math. Mech.*, 1928, 8 (5), 372.

Cooper P. *et al.*, "Performance of Centrifugal Pumps," in *Pumping Station Design*. Oxford: Butterworth Heinemann, 2008.

Dixon, S. L. and C. A. Hall, *Fluid Mechanics and Thermodynamics of Turbomachinery, 7th edition*. Oxford: Elsevier, 2013.

Gülich, J. F. *Centrifugal Pumps*. Berlin: Springer, 2014.

Karassik, I. J.; Krutzsch, W. C.; Fraser, W. H. and Messina, J. P. *Pump Handbook*. New York: McGraw Hill, 1976.

Kothandaraman, R. and Rudramoorthy, C. P. *Fluid Mechanics and Machinery*, second edition. New Delhi: New Age International, 2007.

Krouza, V. *Centrifugal Pumps and Accessories*. Prague: ČSAV, 1956.

Lobanoff, V. S. and Ross, R. R. *Centrifugal Pumps: Design and Application*. Houston, TX: Gulf Professional Publishing, 2013.

Lomakin, A. A. *Centrifugal and Axial Pumps* (in Russian). Leningrad: Mashinostroenie Publ., 1966.

Mikhailov, A. K. and Maliushenko, V. V. *Lobe Pumps. Theory, Calculation and Design Engineering* (in Russian). Moscow: Mashinostroenie Publ., 1977.

Nechleba, M. and Hušek, J. *Hydraulic Machines (in Czech)*. Prague: SNTL / SVTL, 1966.

Neumann, B. *The Interaction Between Geometry and Performance of a Centrifugal Pump*. London: Mechanical Engineering Publications, 1991.

Paciga, G. M. and Strýček, O. *Pumping Techniques (in Slovak)*. Bratislava, Slovakia: ALFA, 1984.

Rama, S. R. G. and Khan, A. *Turbomachinery Design and Theory*. New York: Marcel Dekker, 2003.

Strýček, O., *Hydrodynamic Pumps*. Bratislava: Slovakia: STU Bratislava, 1994.

Steponoff, A. J. *Centrifugal and Axial Flow Pumps: Theory, Design and Application*, second edition. Malabarf, FL: Krieger, 1957.

Stepanoff, A. J. *Centrifugal and Axial Flow Pumps: Theory, Design, and Application*. Malabar, FL: Krieger, 1991.

Stodola, A. *Steam and Gas Turbines with a Supplement on The Prospects of the Thermal Prime Mover*. New York: McGraw-Hill, 1927.

Sultanian, B. K. *Logan's Turbomachinery*. Boca Raton: CRC Press, 2019.

Turton, R. K. *Principles of Turbomachinery*, London: Chapman and Hall, 1984.

Varchola, M. "Hydro turbine for small head and high speed," in *Hydroturbo 2010: 20th International Conference "On the Use of Hydropower"*, Slovakia: Slovak Technical University (STU), 2010.

Varchola, M. and Hlobocan, P., *Hydraulic Design of Centrifugal Pumps*. Bratislava: Slovakia: STU Bratislava, 2016.

Xu, L.; Ji, D.; Shi, W.; Xu, B.; Lu, W. and Lu, L. "Influence of inlet angle of guide vane on hydraulic performance of an axial flow pump based on CFD," *Shock Vib.*, 2020, 8880789.

4 Axial Flow Pumps

4.1 INTRODUCTION

Axial flow pumps rotate the impeller to move water ahead axially. Pumping huge amounts of water against low heads is best done with axial flow pumps. The suction capability of axial flow pumps is limited; thus, they must be started under water. They are commonly used for land drainage, irrigation, or transporting large amounts of water from a river to a ground-level storage facility. Axial flow pumps are made up of a casing and a propeller-like impeller. An axial flow pump's compact design, as well as its ability to operate at extremely high speeds, are both advantages.

Axial flow pumps have the highest flow rates and lowest discharge pressures compared to radial and mixed-flow pumps. Flow in axial pumps is in a parallel direction to the impeller shaft. The impeller is like a propeller and has a small number of vanes, usually three or four. Axial flow pumps can be equipped with exit guide vanes (Figure 4.1a) to help minimize whirl as well as convert the kinetic energy of the fluid to pressure. The axial flow pump's guiding vane is utilized to recover the flow's kinetic energy at the impeller exit. For many applications, such as drainage and irrigation, high flow rate at low head is required. This makes axial flow pumps an ideal fit. Figure 4.1b shows an elbow-type axial flow pump.

4.2 THEORETICAL AND ACTUAL HEAD

We start by writing Euler's turbomachinery equations for pumps. As discussed in the previous chapters, the Euler equation is based on the conservation of angular momentum and conservation of energy principles. Euler's equation describing shaft torque is given by

$$M_k = \rho Q\left(c_{u2}r_2 - c_{u1}r_1\right) \tag{4.1}$$

where

- M_k is the shaft torque
- ρ is fluid density
- Q is flow rate
- c_{u2} is the tangential component of the absolute velocity at exit
- r_2 is exit radius
- r_1 is inlet radius
- c_{u1} is the tangential component of the absolute velocity at the inlet.

DOI: 10.1201/9781003007142-4

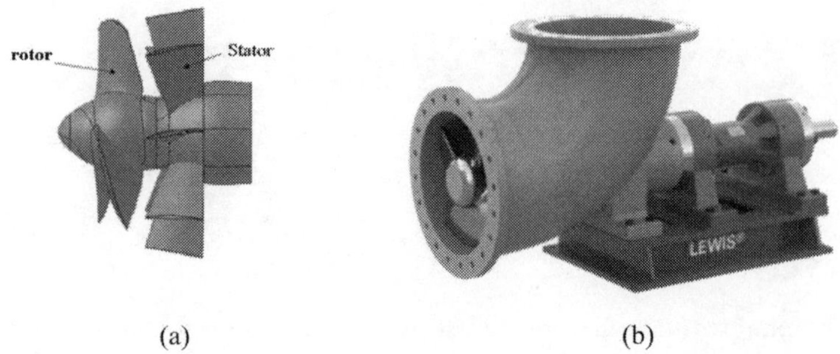

FIGURE 4.1 (a) Rotor with stationary guide vane; (b) elbow-type axial pump (www.global.weir/).

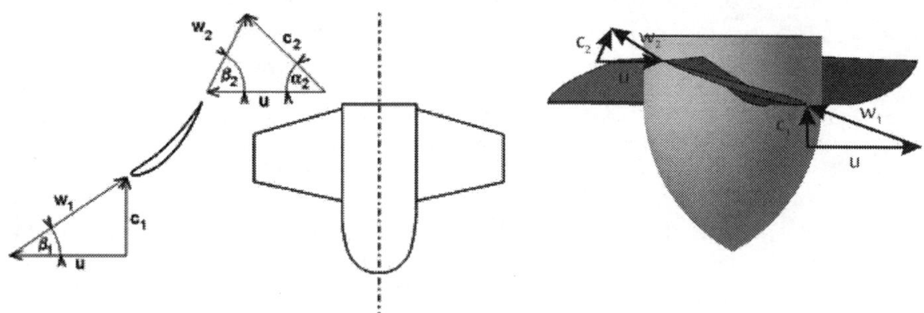

FIGURE 4.2 Velocity triangles of an axial pump.

Power is given by:

$$P = M_k \varpi = \rho Q\left(c_{u2}u_2 - c_{u1}u_1\right) = \rho QgH \tag{4.2}$$

The theoretical head, H_t, is given by

$$H_t = \frac{P}{\rho Qg} = \left(c_{u2}u_2 - c_{u1}u_1\right) \tag{4.3}$$

Velocity triangles at the inlet and outlet of an axial flow pump blade are shown in Figure 4.2.

4.2.1 Axial Flow Cascade

The cascade view of an axial flow pump is given in Figure 4.3. The blades of axial flow pumps feature profiles that are similar to those of an airfoil: they are thin,

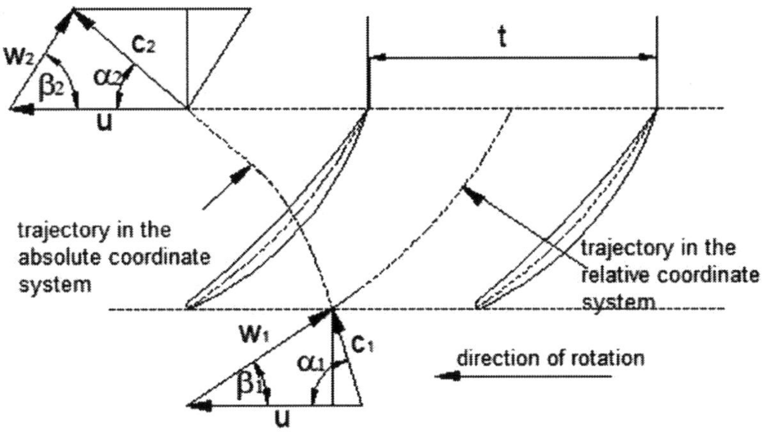

FIGURE 4.3 Flow in a cascade of an axial flow pump.

streamlined, and cambered (Figure 4.3). The relative velocity W_1 approaches the blade at an inlet flow angle β_1, as shown in Figure 4.3. The fluid leaves at outlet flow angle β_2 with relative velocity W_2, which has been slightly deflected. That is, the impeller blades are designed such that the blade angle at exit, β_2, is slightly greater than the vane angle at inlet β_1. This results in a lift force F_L perpendicular to the mean direction of W_1 and W_2, that is, perpendicular to a mean relative velocity, W_m. The pitch t at any radius r is given by $t = 2\pi r/z$, where r is the radius and z is the number of blades.

4.3 FLOW OVER ISOLATED AIRFOILS

Airfoils have found widespread usage in the field of axial flow pumps. Hence, familiarity with airfoil features is required. An airfoil is designed in such a way that the mean line curvature changes the flow direction. The airfoil submerged in a flowing fluid is subjected to viscous force and unbalanced pressure distribution. Due to both viscous force and an unbalanced pressure distribution around the airfoil, the airfoil will be subjected to force. Using the following figure (Figure 4.4), we will describe the terminology and geometry of an airfoil.

The geometry of many airfoil sections is uniquely defined by the National Advisory Committee for Aeronautics (NACA) designation for the airfoil. NACA identified and developed airfoil shapes in a logical manner in the 1930s. NACA airfoil classifications include NACA four-digit wing sections, NACA five-digit wing sections, and NACA six-series wing sections. NACA four-digit wing sections are designated as: NACA ABCDD. The first integer, A, indicates the maximum value of the mean camber line ordinate in percent of chord. The second integer, B, indicates the distance from the leading edge to the maximum camber location in tenths of

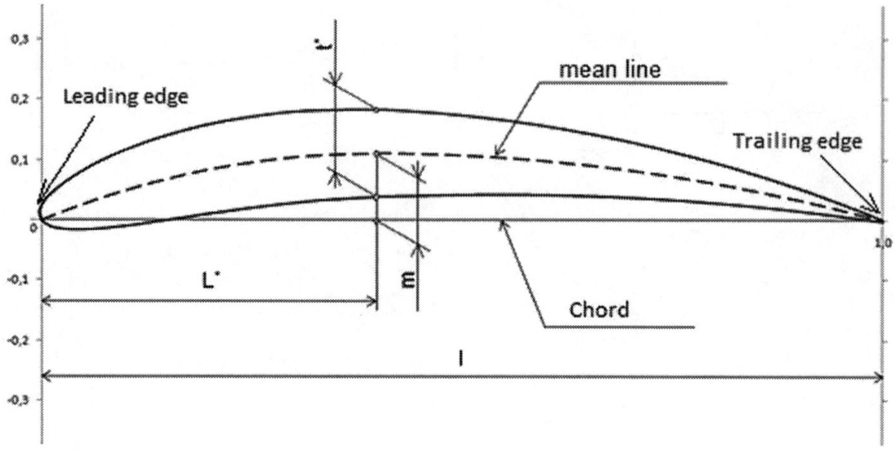

FIGURE 4.4 Airfoil notation according to National Advisory Committee for Aeronautics (NACA).

m is maximum camber
L^* is location of maximum camber
l is length of chord
L is span
t^* is thickness at the maximum camber location
$\frac{m}{l}$ is ratio of maximum camber to chord length
$\frac{L^*}{l}$ is ratio of distance of maximum camber to chord length
$\frac{t^*}{l}$ is ratio of thickness at maximum camber to chord length.

chord. The last two integers, DD, indicate the maximum section thickness in percent of chord. NACA airfoil profiles can be generated with this four-digit information. An online tool, AeroToolbox (https://aerotoolbox.com/naca-4-series-airfoil-generator/), can be used to calculate, plot, and extract airfoil coordinates for any NACA four-digit series airfoil.

4.3.1 Aerodynamics Characteristics of an Airfoil Profile

When an airfoil is exposed to a flow of air, the forces acting on it may be broken down into two parts. These forces can be expressed in terms of nondimensional parameters, lift, and drag coefficient, C_L and C_D (Figure 4.5)

$$F_L = C_L bl\rho \frac{w_m^2}{2} \tag{4.4}$$

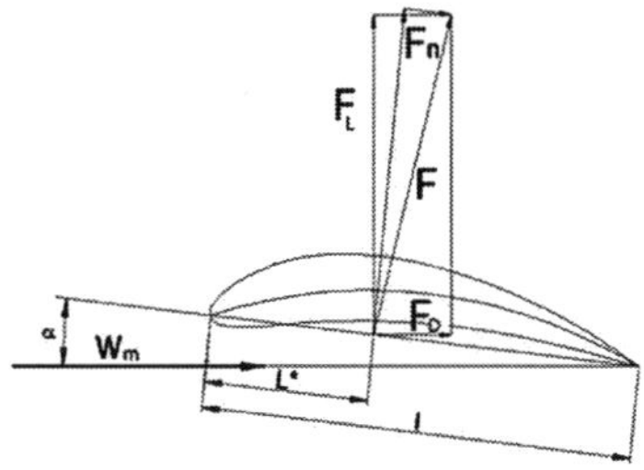

FIGURE 4.5 Aerodynamic forces acting on an airfoil.

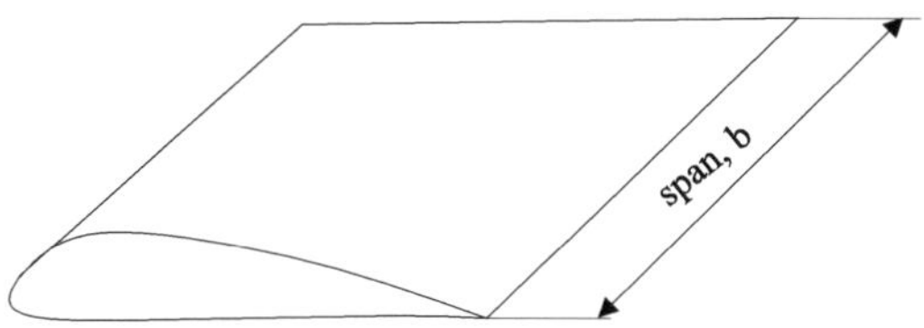

FIGURE 4.6 Airfoil nomenclature, span.

$$F_D = C_D b l \rho \frac{w_m^2}{2} \tag{4.5}$$

where

- b is the span/width of the airfoil (Figure 4.6)
- l is the chord length
- W_m is undisturbed mean relative air velocity
- ρ is density of the fluid
- C_L is the lift coefficient
- C_D is the drag coefficient.

The pitching moment (torque) is given by:

$$\tau_D = C_M b l^2 \rho \frac{w_m^2}{2} \tag{4.6}$$

It is noted that the area $A = bl$ is the maximum projected area of the airfoil. The lift and drag coefficients are affected by the airfoil profile, the angle of attack α, and the aspect ratio. Their values have been determined experimentally for a great number of profiles. The ratio of the length of the airfoil to the length of the chord is called the aspect ratio. The value of C_D is small compared to the value of C_L (Figure 4.7). The angle of attack is the angle formed by the chord line and the fluid direction, and it has a significant impact on the drag coefficient.

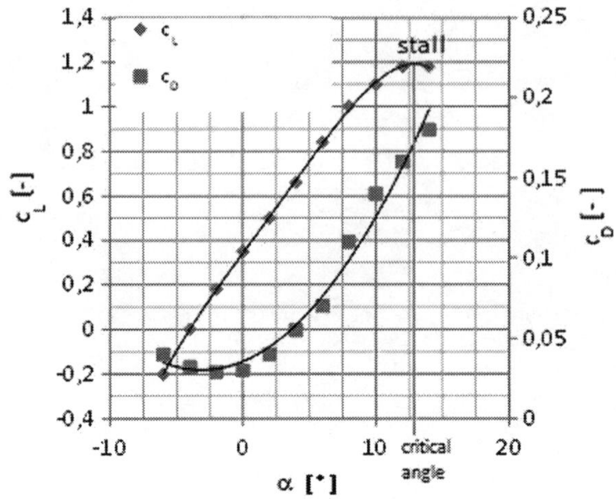

FIGURE 4.7 Lift and drag coefficients for a typical airfoil as a function of attack angle, α.

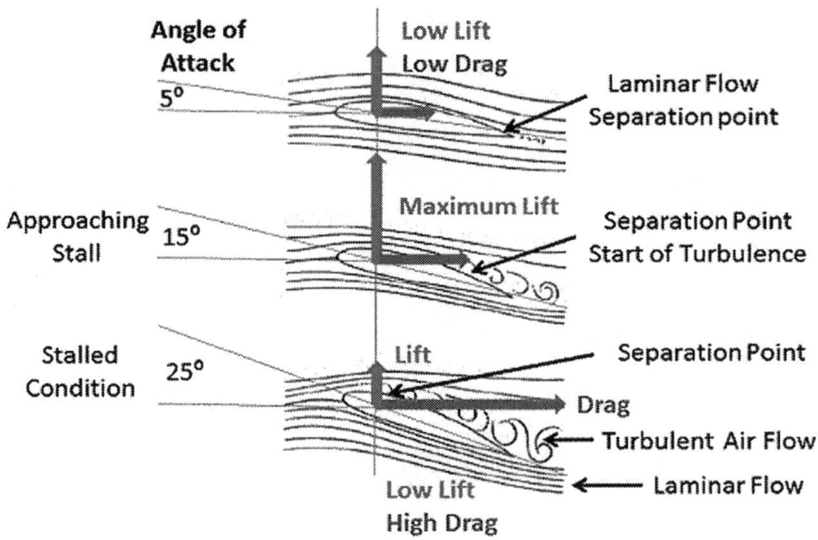

FIGURE 4.8 Development of lift and drag forces at different angles of attack.

Figure 4.8 illustrates development of lift and drag forces at different angles of attacks. The airflow over the airfoil is smooth and laminar at low angles of attack, with a minor amount of turbulence occurring near the trailing edge of the airfoil. The area of the airfoil facing directly into the fluid grows as the angle of attack rises. This improves lift (Figure 4.8), as well as shifting the separation point of laminar flow above the airfoil part towards the leading edge. This results in higher drag because of the higher turbulent flow above the airfoil (Figure 4.8).

4.4 PRESSURE RISE ON AN AIRFOIL

Consider a cascade with the inlet and exit flow angles shown in Figure 4.9. Assume a control volume surrounding a single moving blade, of width S and of unit height along the blade. Assuming that the blade blockage effect on the fluid is negligible the axial component of fluid velocity does not change from inlet to outlet.

Assuming also that the mass flow rate remains constant, we can write the axial force as follows from the momentum equilibrium:

$$F_a = (p_2 - p_1)s = F_L \sin\beta_m - F_D \cos\beta_m \tag{4.7}$$

The pressure rise from inlet to outlet, Δp, in terms of the lift and drag coefficients, is given by

$$\Delta p = p_2 - p_1 = \left(\frac{\rho W_m^2}{2}\frac{c}{s}\right)(C_L \sin\beta_m - C_D \cos\beta_m) \tag{4.8}$$

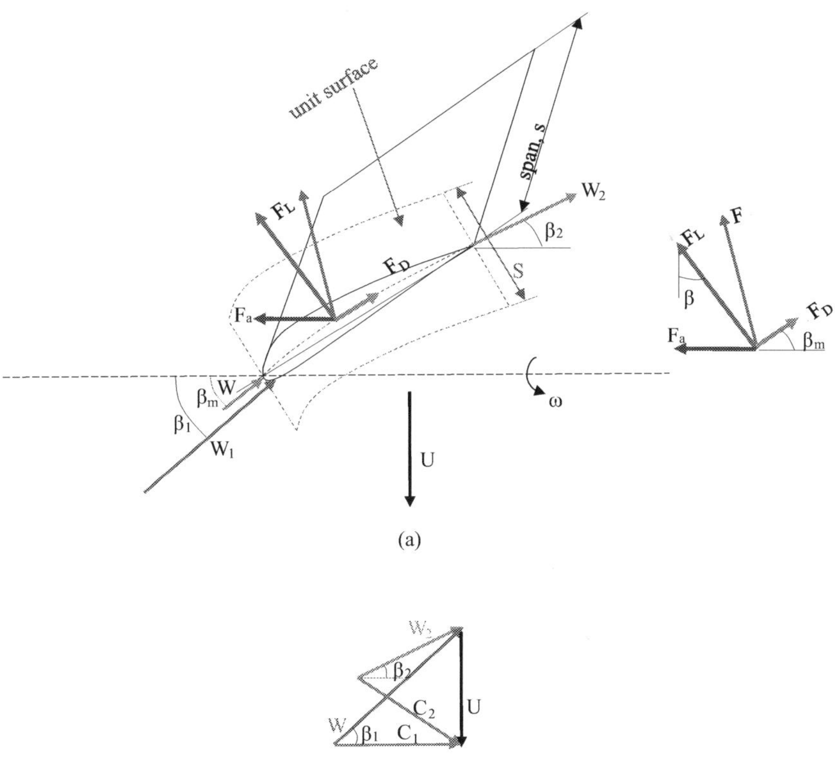

FIGURE 4.9 (a) Control volume for a cascade and (b) velocity triangles.

Neglecting drag coefficient the lift coefficient can be obtained from the following equation:

$$C_L = 2\left(\frac{C}{s}\right)(\tan\beta_1 - \tan\beta_2)\cos\beta_m \tag{4.9}$$

where the span s is given by:

$$s = \frac{2\pi r_m}{z} \tag{4.10}$$

where r_m is the mean radius and z is the number of blades.

4.5 RELATIONSHIP BETWEEN CIRCULATION AND LIFT FORCE

The Kutta–Zhukovsky (Joukowski) theorem relates the lift generated by an airfoil to the speed of the airfoil through the fluid, the density of the fluid, and the circulation around the airfoil. To illustrate the creation of circulation, we will use the figure below. In Figure 4.10, on the left side airflow around an airfoil is shown. In the middle figure circulation around an airfoil due to the vortex is shown. The vector addition of the two flows gives the resultant circulation, shown on the right-hand side of Figure 4.10.

Consider Figure 4.11, in which circulation, Γ, around a single blade of an axial flow machine is shown.

The circulation around one blade in a cascade can be determined for a pump as follows:

$$\Gamma_z = \frac{\Gamma}{z} = \frac{2\pi r(c_{u2} - c_{u1})}{z} = s(c_{u2} - c_{u1}) \tag{4.11}$$

where:

z is the number of blades
s is span.

Circulation around the blade creates a lift force on the blade. To find out what relative velocity, w, is decisive in Kutta–Joukowski theorem for lift, we calculate the force on the blade from the momentum equation, choosing the unit surface area bound by corners **"abcd,"** as indicated in Figure 4.12.

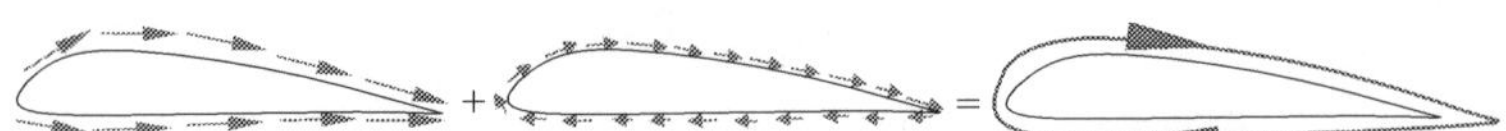

FIGURE 4.10 Creation of circulation around an airfoil.

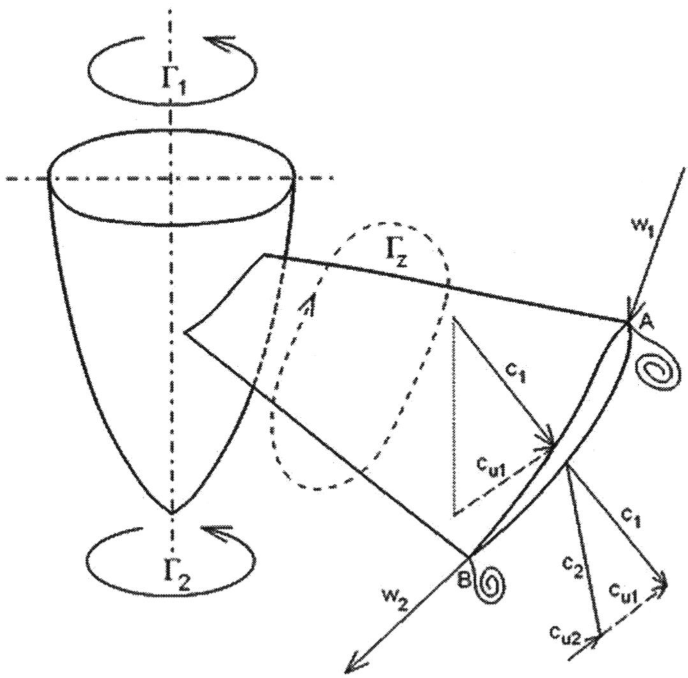

FIGURE 4.11 Circulation, Γ_z, around a blade of an axial machine.

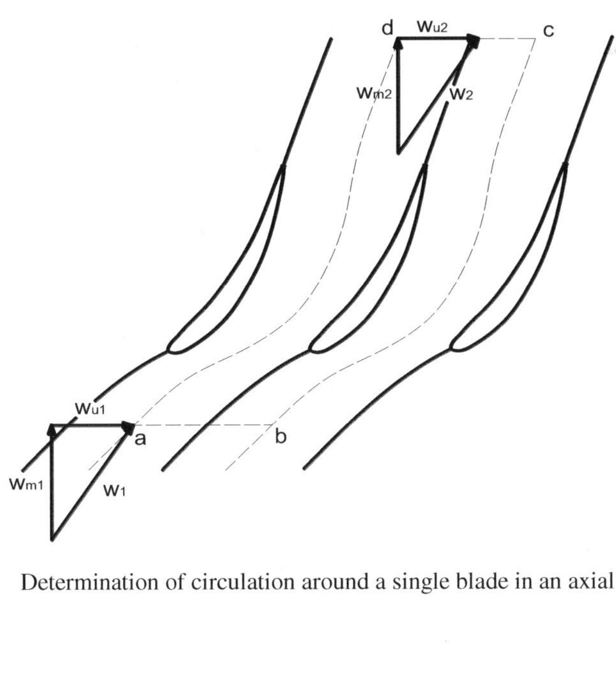

FIGURE 4.12 Determination of circulation around a single blade in an axial cascade.

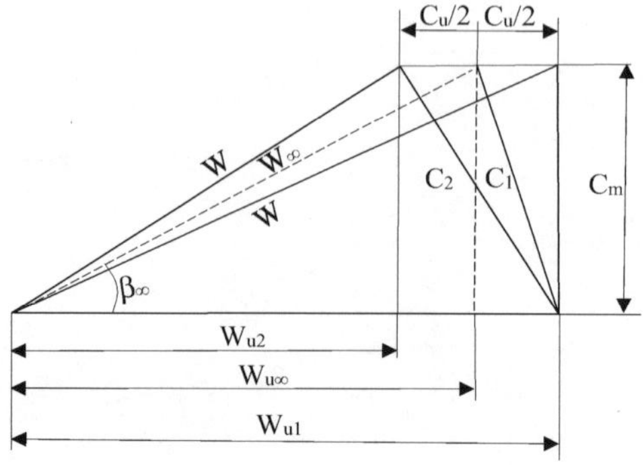

FIGURE 4.13 Velocity triangles at the inlet and outlet.

In the circumferential direction, forces due to the pressure and the forces due to internal friction of the liquid cancel each other. The magnitude of the relative speed changes in the circumferential direction from W_{u1} at the inlet to W_{u2} at exit (Figure 4.13). The circumferential component of the force F_u acting on one blade is then given by:

$$F_u = s w_m \rho \left(W_{u1} - W_{u2}\right) \tag{4.12}$$

where w_m is the relative velocity component in the meridional direction.

The force in the axial direction can be written as:

$$F_a = s\left(p_{u1} - p_{u2}\right) \tag{4.13}$$

The Bernoulli equation can be written for two points (from inlet to exit) and denoting the losses in the control surface as e_z, we arrive at:

$$\frac{p_1 - p_2}{\rho} = \frac{w_2^2 - w_1^2}{2} + e_z = \frac{w_{u2}^2 - w_{u1}^2}{2} + e_z = \frac{w_{u2}^2 - w_{u1}^2}{2}\left(1 + \frac{2\,e_z}{w_{u2}^2 - w_{u1}^2}\right) \tag{4.14}$$

The resultant force acting on the blade is the vector sum of the axial and circumferential forces given by:

$$F_L = \sqrt{F_u^2 + F_a^2} \tag{4.15}$$

Using Equations (4.12–4.15), the lift force is given by:

$$F_L = \sqrt{s^2 \rho^2 \left\{ w_m^2 \left(w_{u1} - w_{u2}\right)^2 + \left(w_{u2} - w_{u1}\right)^2 \left(\frac{w_{u2} + w_{u1}}{2}\right)^2 \left(1 + \frac{2\,e_z}{w_{u2}^2 - w_{u1}^2}\right)^2 \right\}} \tag{4.16}$$

Simplifying the above equation, we get:

$$F_L = \sqrt{s^2\rho^2\left\{w_m^2\left(w_{u1}-w_{u2}\right)^2+\left(w_{u2}-w_{u1}\right)^2 w_\infty^2\left[1+\left(\frac{2\,e_z}{w_{u2}^2-w_{u1}^2}\right)^2\right]\right\}} \tag{4.17}$$

$$F_L = \rho s\left(w_{u2}-w_{u1}\right)\sqrt{w_m^2+\left\{\left(\frac{w_{u2}+w_{u1}}{2}\right)^2\left[1+\left(\frac{2e_z}{\left(w_{u2}^2-w_{u1}^2\right)}\right)^2\right]\right\}} \tag{4.18}$$

Finally, for the lift force we can write:

$$F_L = \rho\Gamma w_\infty \tag{4.19}$$

where we have denoted $s\left(w_{u2}-w_{u1}\right)=\Gamma$

Over the span b of the blade, the lift force is given by

$$F_L = \rho\Gamma w_\infty b \tag{4.20}$$

For an ideal flow with no losses in the control surface, we can easily describe the mean relative velocity, w_∞, as shown in Figure 4.14. Its direction relative to the circumferential direction is given by β_∞. For real flow, there are losses given by e_z. As seen in Equation (4.15) the magnitude of w_∞ will increase by $\dfrac{e_z}{w_{u2}-w_{u1}}$, resulting in w'_∞, which is greater than w_∞. Its direction is given by β'_∞. The meridional velocity component, w_m, however does not change. Therefore, the resulting lift force over the span of the blade is given by:

$$F'_L = \rho\Gamma w'_\infty L \tag{4.21}$$

Figure 4.14 shows that the lift force has been reduced from F_L to F_L' because of the drag force F_D.

Similarity of triangles for the hatched areas in Figure 4.14 yields:

$$\frac{F'_L}{F_L-\dfrac{F_D}{tg\beta_\infty}} = \frac{w'_\infty}{w_\infty} \tag{4.22}$$

In general form, drag force and lift force are given by $F_D = C_D\rho\dfrac{v^2}{2}A$ and $F_L = C_L\rho\dfrac{v^2}{2}A$. Therefore, we can write:

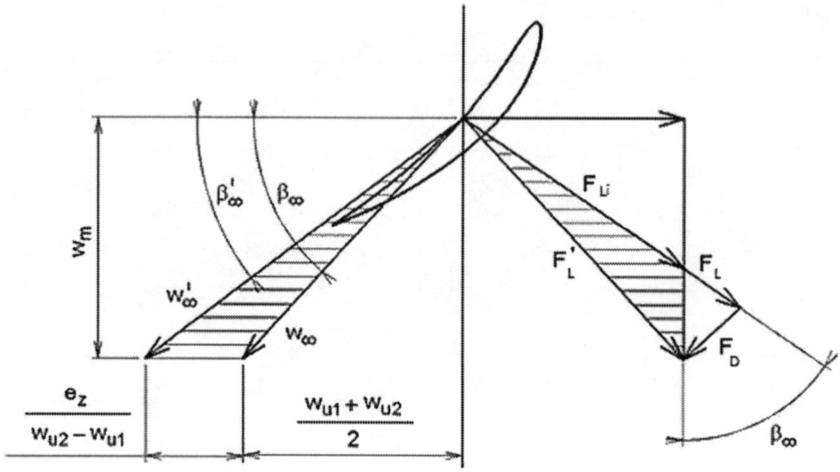

FIGURE 4.14 Velocity triangles accounting for losses in the control surface.

$$F_L - \frac{F_D}{tg\beta_\infty} = \rho\frac{C_L}{2}Llw_\infty^2 - \rho\frac{C_D}{2}Ll\frac{w_\infty^2}{tg\beta_\infty} \tag{4.23}$$

where l is streamline length in the meridional section.

Equation 4.23 can be further simplified and rewritten as follows:

$$F_L - \frac{F_D}{tg\beta_\infty} = \rho Ll\frac{w_\infty^2}{2}\left(C_L - \frac{C_D}{tg\beta_\infty}\right) \tag{4.24}$$

Using Equation 4.17 and the geometry shown in Figure 4.14, we can further rewrite Equation 4.24 in the following form:

$$\rho L\Gamma w'_\infty w_\infty = \rho C_L Ll\frac{w_\infty^2}{2}w'_\infty - \rho\frac{C_D}{2}Ll\frac{w_\infty^2}{tg\beta_\infty}w'_\infty \tag{4.25}$$

which finally gives us the circulation in terms of lift and drag coefficients, in the following form:

$$\Gamma = \frac{1}{2}\left(C_L - \frac{C_D}{tg\beta_\infty}\right)lw_\infty \tag{4.26}$$

Equation 4.26 is important because it shows that, to determine circulation, we really do not need to know w'_∞ and β'_∞. All we need to know is w_∞ and β_∞, which can be determined from a much simpler velocity triangle, as shown in Figure 4.14.

Using Equations 4.8 and 4.25, we can write

$$s(c_{u1} - c_{u2}) = \frac{1}{2}\left(C_L - \frac{C_D}{tg\beta_\infty}\right) l w_\infty \tag{4.27}$$

which yields:

$$C_L - \frac{C_D}{tg\beta_\infty} = 2\frac{c_{u1} - c_{u2}}{w_\infty}\frac{s}{l} = \frac{2\Gamma}{l w_\infty} \tag{4.28}$$

Using Euler's turbomachinery equation, $u(c_{u2} - c_{u1}) = \frac{gH}{\eta_h}$, Equation 4.28 can be rewritten as

$$C_L - \frac{C_D}{tg\beta_\infty} = \frac{gH}{\eta_h u w_\infty}\frac{s}{l} \tag{4.29}$$

Finally, the hydraulic efficiency is given by

$$\eta_h = \frac{gH}{u w_\infty}\frac{t}{l}\frac{1}{c_L - \frac{c_D}{tg\beta_\infty}} \tag{4.30}$$

Equation 4.30 shows that, for very high specific speed, i.e., for very high values of u and w_∞, l should be shorter to maintain high hydraulic efficiency. Another way to maintain high hydraulic efficiency is to increase the span, s. Another possible way to maintain higher efficiency for very high values of u and w_∞ is to decrease the number of blades.

Equation 4.30 can be used to determine suitable blade profiles with respect to the selected specific speed (or w_∞) of an axial pump. If the lift and drag coefficients in Equation 4.30 are satisfied, then the selected profile will have such properties that circulation is created around the blade, which is necessary to achieve the required pressure. Lift and drag coefficients described above are for an isolated blade in cascade. In practice since there is more than one blade in a cascade, it is necessary to introduce some correction factors.

4.6 PRELIMINARY DESIGN OF IMPELLER OF AN AXIAL FLOW MACHINE

4.6.1 Efficiency and Power

We assume that the flow rate and the head are known. First, we calculate the specific speed:

$$n_b = n\frac{\sqrt{Q}}{(gH)^{0.75}} \tag{4.31}$$

$$n_s = 1213.9 n_b \tag{4.32}$$

Next, we calculate the overall efficiency, hydraulic efficiency, and power using the following equations.

The overall efficiency for $0.04 \le n_b \le 0.33$

$$\eta_c = \sqrt{\left(\frac{Q}{0.048}\right)^{0.083} - \left|-0.722 - \log n_b\right|^3} - \xi \qquad (4.33)$$

$$\text{For } n_b > 0.33 \quad \eta_c = \sqrt{\left(\frac{Q}{0.048}\right)^{0,7} - \left|-0.722 - \log n_b\right|^5} - \xi \qquad (4.34)$$

$\xi = 0.2$ for single-stage pump. Note that axial flow pumps are mostly single-stage pumps.

$\xi = 0.22$ for multi-stage pumps

The above equations apply for flow rates up to 0.65 m^3s^{-1}. For flow rates Q= 0.65 m^3s^{-1} and above the following equations are used.

$$\text{For } 0.04 \le n_b \le 0.33 \quad \eta_c = \sqrt{1,24 - \left|-0,722 - \log n_b\right|^3} - \xi \qquad (4.35)$$

$$\text{For } n_b > 0.33 \quad \eta_c = \sqrt{1.24 - \left|-0.722 - \log n_b\right|^5} - \xi \qquad (4.36)$$

Another relationship for calculating the overall efficiency for flow rates Q > 0.178 m^3s^{-1} (Mikhailov and Maliushenko 1977).

$$\eta_c = 1 - \frac{0.06 + 0.304 n_b + \dfrac{43.2}{(332.6 n_b + 8.2)^2} + \dfrac{0.178 - Q}{2.13 + 66Q}}{Q^{\frac{1}{8}}} \qquad (4.37)$$

Once the overall efficiency is calculated, the hydraulic efficiency is determined from the following equation:

$$\eta_h = \sqrt{\eta_c} - (0.02 \text{ } to \text{ } 0.06) \qquad (4.38)$$

Finally, the power is obtained as:

$$P = \frac{\rho Q g H}{\eta_h} \qquad (4.39)$$

4.7 MERIDIONAL SECTION OF AN AXIAL FLOW MACHINE

We start with the determination of velocity in the meridional section.

$$c_m = K_m \sqrt{2gH_n} \tag{4.40}$$

The constant K_m is calculated as follows:

$$K_m = 0.0688 + 0.733 n_b^{1,1} \tag{4.41}$$

$$c_m = \frac{Q + \sum Q_z}{\frac{\pi}{4}\left(D_n^2 - D_A^2\right)} \tag{4.42}$$

ΣQ_z accounts for leakage and is usually taken as 4%. The diameters at different sections of the axial flow machine are calculated as follows (Figure 4.15).

$$\frac{D_B}{D_A} = 0.63 - 0.346(n_b - 0.25) \tag{4.43}$$

where

$D_B = D_n$ is the hub diameter
$D_A = D_2$ is the outlet (shroud) diameter

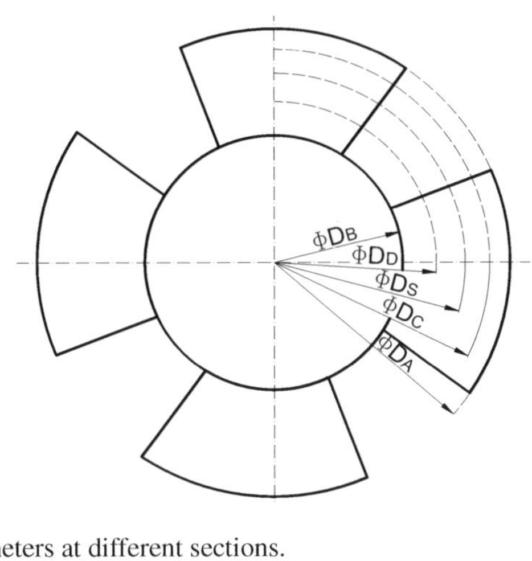

FIGURE 4.15 Diameters at different sections.

$$D_A = \sqrt{\frac{4}{\pi}\frac{\left(Q+\sum Q_z\right)}{c_m}\frac{1}{1-\left(\frac{D_n}{D_A}\right)^2}} \tag{4.44}$$

$$D_S = \sqrt{\frac{D_A{}^2 + D_n{}^2}{2}} \tag{4.45}$$

$$D_C = \sqrt{\frac{D_S{}^2 + D_A{}^2}{2}} \tag{4.46}$$

$$D_D = \sqrt{\frac{D_B{}^2 + D_S{}^2}{2}} \tag{4.47}$$

4.8 NUMBER OF BLADES

The number of blades can be determined from Figure 4.16, which shows the relationship between specific speed, n_b, and central angle. The central angle ϕ_r (Figure 4.17) depends on specific speed, D_B/D_A ratio, and number of blades z (Figure 4.15). When selecting the number of blades Equation 4.43 should also be satisfied.

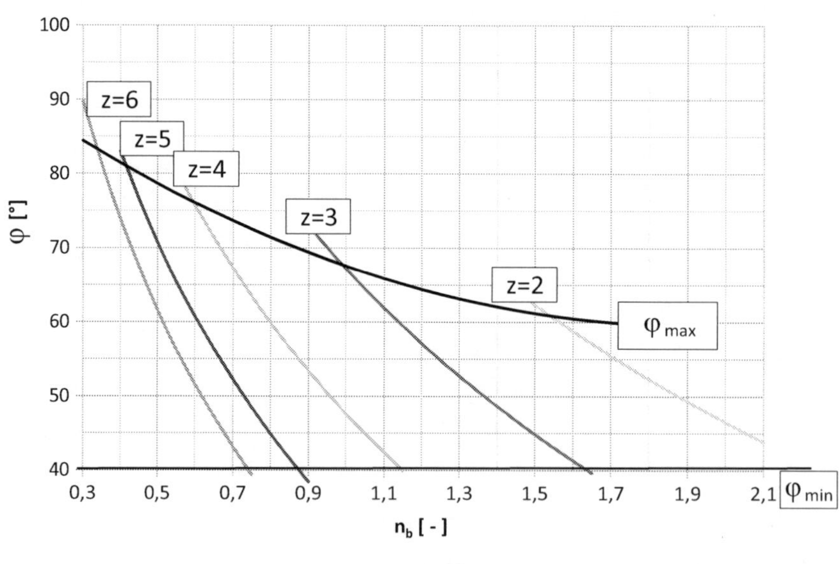

FIGURE 4.16 Nomogram for determining the number of blades and the center angle based on the specific speed, n_b.

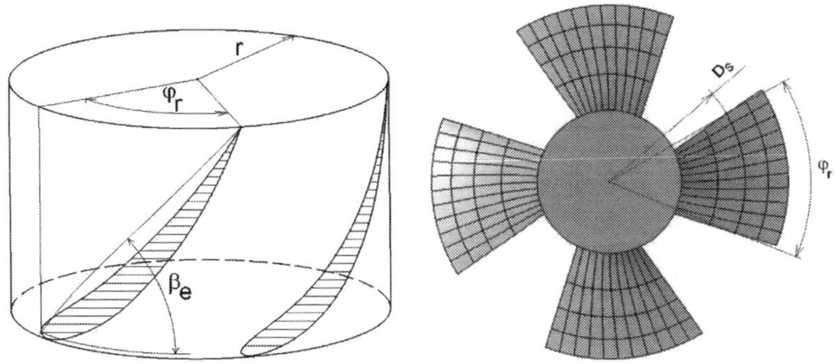

FIGURE 4.17 View of the cylindrical surface of the blade section.

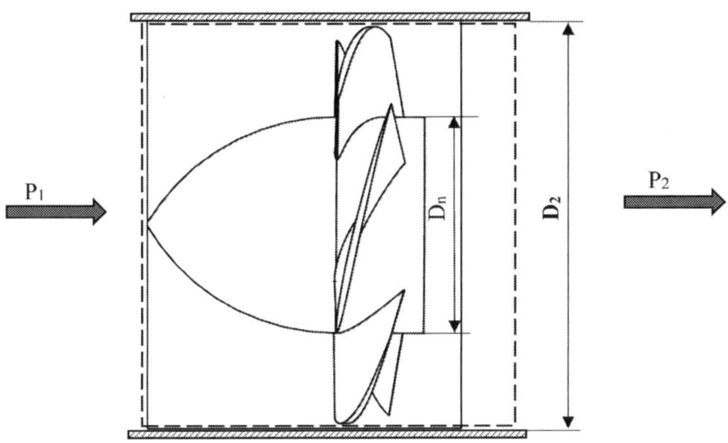

FIGURE 4.18 Pressure differential in a control volume of an axial pump.

4.9 AXIAL THRUST IN AXIAL PUMPS

In axial impellers, any pressure differential between the two sides of the vanes is translated into only axial thrust on the rotating element (Figure 4.18). There is no radial thrust in axial pumps.

In axial impellers, the axial force arises mainly due to the pressure difference between the pressure and suction side of the vanes (Figure 4.19). This pressure difference causes an axial force. The area on which the pressure difference of the axial pump acts is given by the area of the annulus between the hub and the casing. The axial force can be calculated using Equation (4.51):

$$F_A = \pi\rho\left(r_2^2 - r_1^2\right)1.3Y + \pi\rho\left(r_1^2 - r_n^2\right)\left[1.3Y - \frac{\varpi^2}{16}\left(r_1^2 - r_n^2\right)\right] \tag{4.48}$$

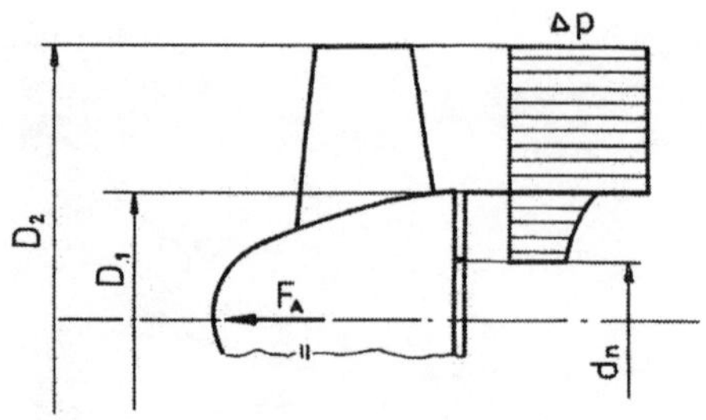

FIGURE 4.19 Axial force in an axial pump.

$$F_A = \pi\rho\left(D_2^2 - D_1^2\right)\frac{1.3}{4}gH + \pi\rho\frac{\left(D_1^2 - D_n^2\right)}{4}\left[1.3gH - \frac{\varpi^2}{(16)(4)}\left(D_1^2 - D_n^2\right)\right] \tag{4.49}$$

$$F_A = \pi\rho\left(D_2^2 - D_1^2\right)\frac{1.3}{4}gH + \pi\rho\left(D_1^2 - D_n^2\right)\left[\frac{1.3}{4}gH - \frac{\varpi^2}{256}\left(D_1^2 - D_n^2\right)\right] \tag{4.50}$$

$$F_A = \pi\rho\left(D_2^2 - D_1^2\right)(0.325)gH + \pi\rho\left(D_1^2 - D_n^2\right)\left[(0.325)gH - \frac{\varpi^2}{256}\left(D_1^2 - D_n^2\right)\right] \tag{4.51}$$

where:

F_A is the axial thrust
ρ is density of the fluid
ω is rotational speed in rad/s
H is head.

4.10 PERFORMANCE CHARACTERISTICS OF AXIAL FLOW PUMPS

Axial flow pumps have performance curves that differ from radial flow pumps. Figure 4.20 compares nondimensional performance curves of a radial flow pump (a) and an axial flow pump (b). The figure displays the head, brake horsepower, and efficiency as a percentage of these values at the pump's design or best-efficiency point.

Flow separations in the impeller of an axial flow pump may occur when operating at part load due to a rise in the incidence angle at the impeller leading edge, most likely resulting in head curve instability. The unstable region is shown in Figure 4.21. If the pump is operated within the unstable operation region, there will be vibration and noises. Researchers have also reported that the power consumption can surpass the rated brake power highly (Figure 4.20b). Therefore, it is advised that axial pumps be started with the delivery valve fully open.

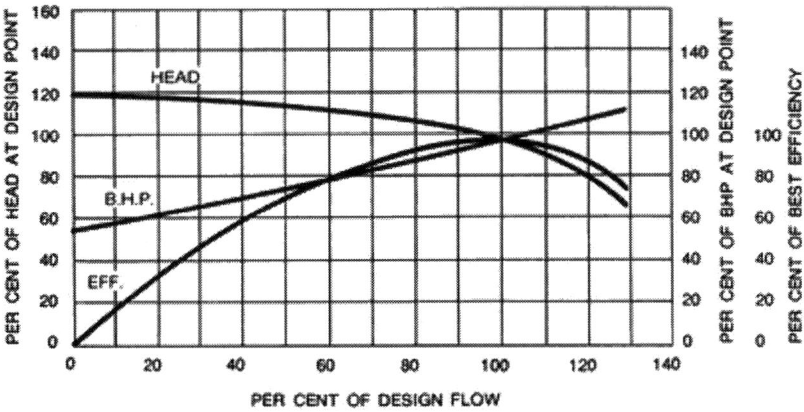

(a)

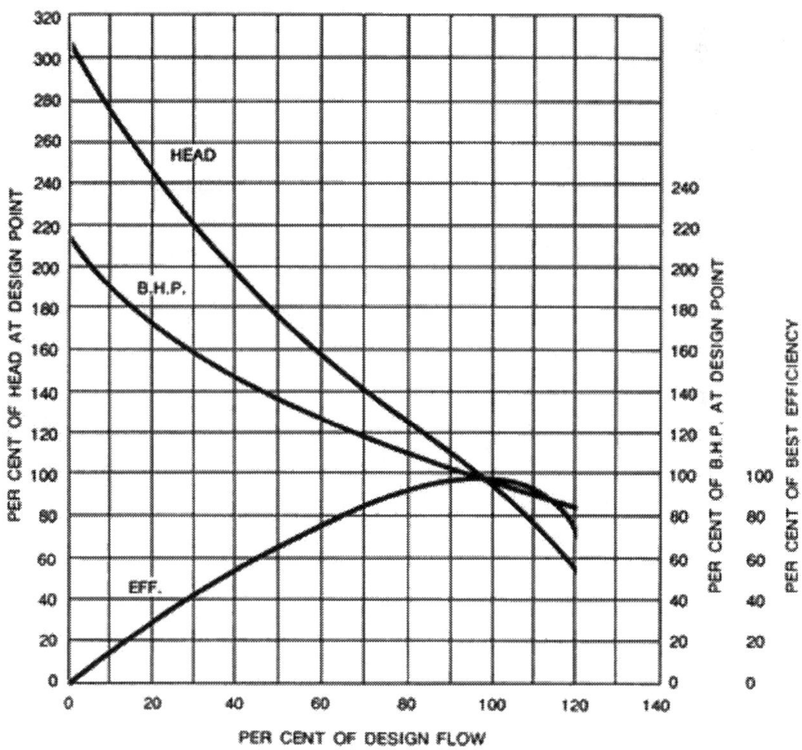

(b)

FIGURE 4.20 Performance characteristics of (a) centrifugal pumps and (b) axial pumps (www.b-k.com/).

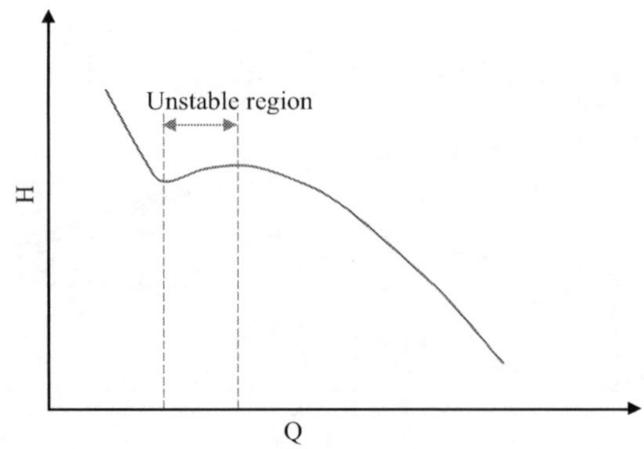

FIGURE 4.21 Unstable region in the head curve of an axial flow pump.

4.11 FLOW RATE CONTROL IN AXIAL PUMPS

Generally, flow in an axial flow pump can be controlled in the following four ways.

1. Using the delivery valve
   Pumps are equipped with throttling delivery valves. The valves can be used to regulate flow rate. From Figure 4.20b, it is clear that reducing flow rate by throttling will result in higher power consumption and reduced efficiency. Hence the use of this method should be carefully considered.
2. Adjusting inlet guide vanes
   If an axial pump's inlet guide blades are adjustable, they can be used to regulate flow. The use of inlet guide blades for flow control will result in reduction of head and input power.
3. Adjusting impeller blade angle
   If the pump has an adjustable impeller blade, the blade orientation can be adjusted to regulate the flow rate. Generally, the decrease in inlet and outlet blade angles will result in reduced head.
4. Controlling pump speed
   If the pump is equipped with a variable speed driver, then the flow rate can be regulated by changing the speed. The flow rate is directly proportional to the pump speed. A higher pump speed will result in higher flow rate, and vice versa.

4.12 EXAMPLE PROBLEMS

Example 4.1

Generate an airfoil profile for the NACA 0010 whose maximum thickness is 10% of the chord using AeroToolbox.

Solution

Figure 4.22 shows the generated profile. Note the symmetry. The two leading "0" digits indicate that there is no camber, so the airfoil is symmetric about the x-axis.

The website also enables extraction of data. These data can be plotted using a spreadsheet or any other suitable software. The 50-point data of the profile are given in Table 4.1.

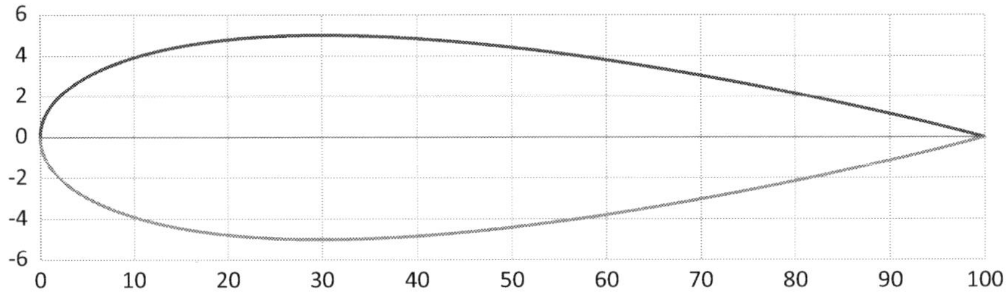

FIGURE 4.22 NACA 0010 airfoil generated with AeroToolbox.

TABLE 4.1
Data of the Airfoil NACA 0010

Upper Surface		Lower Surface		Camber Line	
x	y	x	y	x	y
0	0	0	0	0	0
0.051365	0.234658	0.051365	–0.23466	0.051365	0
0.20525	0.462483	0.20525	–0.46248	0.20525	0
0.461021	0.682945	0.461021	–0.68295	0.461021	0
0.817628	0.895338	0.817628	–0.89534	0.817628	0
1.273606	1.098806	1.273606	–1.09881	1.273606	0
1.827081	1.292374	1.827081	–1.29237	1.827081	0
2.475778	1.474984	2.475778	–1.47498	2.475778	0
3.217032	1.645539	3.217032	–1.64554	3.217032	0
4.047797	1.802946	4.047797	–1.80295	4.047797	0
4.964659	1.946161	4.964659	–1.94616	4.964659	0
5.963851	2.074231	5.963851	–2.07423	5.963851	0
7.041266	2.18633	7.041266	–2.18633	7.041266	0
8.192478	2.281794	8.192478	–2.28179	8.192478	0
9.412755	2.360148	9.412755	–2.36015	9.412755	0
10.69708	2.42112	10.69708	–2.42112	10.69708	0
12.04019	2.464654	12.04019	–2.46465	12.04019	0
13.43654	2.490912	13.43654	–2.49091	13.43654	0
14.88042	2.500261	14.88042	–2.50026	14.88042	0
16.36587	2.493269	16.36587	–2.49327	16.36587	0
17.88681	2.470677	17.88681	–2.47068	17.88681	0
19.43698	2.43338	19.43698	–2.43338	19.43698	0
21.01	2.382398	21.01	–2.3824	21.01	0
22.59942	2.318848	22.59942	–2.31885	22.59942	0
24.19871	2.243915	24.19871	–2.24392	24.19871	0
25.80129	2.158827	25.80129	–2.15883	25.80129	0
27.40058	2.064833	27.40058	–2.06483	27.40058	0
28.99	1.963177	28.99	–1.96318	28.99	0
30.56302	1.855092	30.56302	–1.85509	30.56302	0
32.11319	1.741781	32.11319	–1.74178	32.11319	0
33.63413	1.624417	33.63413	–1.62442	33.63413	0

TABLE 4.1 (Continued)
Data of the Airfoil NACA 0010

Upper Surface		Lower Surface		Camber Line	
x	y	x	y	x	y
35.11958	1.504138	35.11958	–1.50414	35.11958	0
36.56346	1.382048	36.56346	–1.38205	36.56346	0
37.95981	1.259226	37.95981	–1.25923	37.95981	0
39.30292	1.136726	39.30292	–1.13673	39.30292	0
40.58725	1.015584	40.58725	–1.01558	40.58725	0
41.80752	0.896827	41.80752	–0.89683	41.80752	0
42.95873	0.781472	42.95873	–0.78147	42.95873	0
44.03615	0.670528	44.03615	–0.67053	44.03615	0
45.03534	0.564993	45.03534	–0.56499	45.03534	0
45.9522	0.465853	45.9522	–0.46585	45.9522	0
46.78297	0.374065	46.78297	–0.37407	46.78297	0
47.52422	0.290551	47.52422	–0.29055	47.52422	0
48.17292	0.216181	48.17292	–0.21618	48.17292	0
48.72639	0.151757	48.72639	–0.15176	48.72639	0
49.18237	0.097995	49.18237	–0.098	49.18237	0
49.53898	0.055509	49.53898	–0.05551	49.53898	0
49.79475	0.024795	49.79475	–0.0248	49.79475	0
49.94864	0.006217	49.94864	–0.00622	49.94864	0
50	0	50	0	50	0

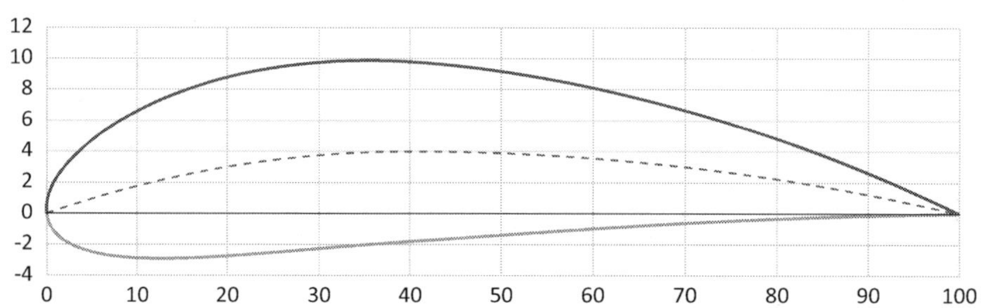

FIGURE 4.23 NACA 4412 airfoil generated with AeroToolbox.

Example 4.2

Generate an airfoil profile for the NACA 4412 airfoil section.

Solution

The designation tells us that this airfoil is 12% thick and has a 4% maximum camber located at 4/10th (40%) of the chord. The profile generated using AeroToolbox is given in Figure 4.23 and Table 4.2.

TABLE 4.2
Data of the Airfoil NACA 4412

Upper Surface		Lower Surface		Camber Line	
x	y	x	y	x	y
0	0	0	0	0	0
−0.00372	0.286408	0.106453	−0.26589	0.051365	0.01026
0.097484	0.585255	0.313015	−0.50358	0.20525	0.040839
0.303864	0.895466	0.618178	−0.71318	0.461021	0.091142
0.615222	1.215351	1.020035	−0.89499	0.817628	0.160183
1.030905	1.54265	1.516308	−1.04943	1.273606	0.246611
1.549788	1.874582	2.104374	−1.17713	1.827081	0.348725
2.170259	2.207921	2.781298	−1.27891	2.475778	0.464508
2.890199	2.539071	3.543866	−1.35575	3.217032	0.59166
3.706975	2.864158	4.38862	−1.40889	4.047797	0.727636
4.617428	3.179129	5.311891	−1.43974	4.964659	0.869693
5.617872	3.479847	6.30983	−1.44998	5.963851	1.014933
6.704101	3.762196	7.378432	−1.44148	7.041266	1.160356
7.8714	4.022175	8.513555	−1.41635	8.192478	1.302912
9.114572	4.255988	9.710938	−1.37689	9.412755	1.439551
10.42796	4.460131	10.9662	−1.32557	10.69708	1.567279
11.80551	4.631467	12.27486	−1.26505	12.04019	1.683207
13.24078	4.767282	13.63231	−1.19807	13.43654	1.784605
14.72701	4.865339	15.03382	−1.12744	14.88042	1.868949
16.25722	4.923915	16.47453	−1.05598	16.36587	1.933966
17.82417	4.941823	17.94945	−0.98648	17.88681	1.977672
19.42054	4.918425	19.45342	−0.9216	19.43698	1.998415
21.02284	4.856582	20.99717	−0.86112	21.01	1.997733
22.63157	4.767416	22.56728	−0.79745	22.59942	1.984984
24.24895	4.653053	24.14847	−0.73141	24.19871	1.960824
25.86806	4.514943	25.73452	−0.66452	25.80129	1.925211
27.48203	4.354752	27.31912	−0.59817	27.40058	1.878292
29.08405	4.174334	28.89595	−0.53354	28.99	1.8204
30.66742	3.975711	30.45863	−0.47161	30.56302	1.75205
32.22555	3.76105	32.00083	−0.41318	32.11319	1.673935
33.75203	3.532644	33.51622	−0.35882	33.63413	1.586912
35.2406	3.2929	34.99857	−0.30891	35.11958	1.491996
36.68522	3.04432	36.4417	−0.26365	36.56346	1.390338
38.08005	2.789492	37.83958	−0.22307	37.95981	1.283211
39.41951	2.531073	39.18632	−0.18708	39.30292	1.171994
40.69829	2.271777	40.4762	−0.15549	40.58725	1.058145
41.91134	2.014355	41.7037	−0.12799	41.80752	0.943182
43.05393	1.761581	42.86354	−0.10426	42.95873	0.828659
44.12162	1.516222	43.95068	−0.08394	44.03615	0.716141
45.11032	1.281015	44.96036	−0.06665	45.03534	0.607182
46.01626	1.058638	45.88815	−0.05205	45.9522	0.503296
46.83603	0.851671	46.72991	−0.03979	46.78297	0.405939
47.56656	0.662564	47.48189	−0.0296	47.52422	0.316483

TABLE 4.2 (Continued)
Data of the Airfoil NACA 4412

Upper Surface		Lower Surface		Camber Line	
x	y	x	y	x	y
48.20515	0.4936	48.14069	–0.02122	48.17292	0.236193
48.74946	0.346852	48.70333	–0.01443	48.72639	0.16621
49.1975	0.224149	49.16725	–0.00909	49.18237	0.107532
49.54765	0.127041	49.53031	–0.00505	49.53898	0.060997
49.79866	0.056769	49.79084	–0.00222	49.79475	0.027273
49.94962	0.014239	49.94765	–0.00055	49.94864	0.006843
50	0	50	0	50	0

Example 4.3

An axial pump is to be designed at a nominal flow rate of 1000 l/s, specific energy (Y_n) of 49 J/kg (i.e., $gH_n = 49$ J/kg), rotating at 16.66 Hz. Determine the main dimensions of the pump.

Solution

Given

$$Q_n = 1000\frac{L}{s} \quad Y_n = 49\frac{J}{kg} \quad H_n = 4.997m \quad n_n = 159rpm \Rightarrow \varpi = 16.66\frac{1}{s}$$

Specific speed

$$n_b = n_n \frac{\sqrt{Q_n}}{\left(g \cdot H_n\right)^{\frac{3}{4}}} = 0.899$$

$$n_s = 1213.9 \cdot n_b = 1091$$

$$For\ Q < 0.65\frac{m^3}{s} \qquad n_b \geq 0.33$$

Overall efficiency, where $\xi = 0.2$ *for a single stage pump*

$$\eta_c = \sqrt{1.24 - \left(\left|-0.722 - \log\left(n_b\right)\right|\right)^5} - \xi = 0.848$$

Hydraulic efficiency will be calculated Wislicenus equation

$$\eta_h = \sqrt{\eta_c} - 0.03 = 0.891 \quad Note\ that\ \eta_h = \sqrt{\eta_c} - (0.2\ to\ 0.6)$$

Power

$$P = \frac{\rho Q_n g H_n}{\eta_h} = 54990W$$

$$\frac{D_B}{D_A} = 0.63 - 0.346(n_b - 0.25) = 0.405$$

$$K_m = 0.0688 + 0.733 n_b^{1.1} = 0.721$$

$$c_m = K_m \sqrt{2 \cdot g \cdot H_n} = 7.136 \frac{m}{s}$$

$$D_A = \sqrt{\frac{4 \cdot 1.04 \cdot Q_n}{\pi \cdot c_m} \frac{1}{1 - \frac{D_B}{D_A}}} = 0.471$$

$$D_B = 0.405 \cdot D_A = 0.191$$

$$D_n = D_B$$

$$D_s = \sqrt{\frac{D_A^2 + D_n^2}{2}} = 0.36m$$

$$D_D = \sqrt{\frac{D_B^2 + D_s^2}{2}} = 0.288m$$

$z = 4$ *Number of blades is selected u* sing *Fig. 4.15.*
It is a function of specfic speed

$$u = \varpi \cdot \frac{D_A}{2} = 3.923 \frac{m}{s}$$

$$W_\infty = u - \frac{g \cdot H_n}{2 \cdot \pi \cdot D_A \cdot \varpi \cdot \eta_h} = 2.808 \frac{m}{s}$$

Example 4.4

An axial flow pump impeller consists of six blades with chord length, $C = 3.2$ in. Hub diameter, $D_1 = 10$ in, and tip diameter, $D_2 = 15$ in. If $C_L = 1.3$ and $C_D = 0.025$ and the impeller is rotating at 1500 rpm, determine the flow rate and static pressure across the impeller. Assume inlet blade angle $\beta_1 = 45°$ and outlet blade angle $\beta_2 = 22°$.

$N := 1500\,rpm$ $\quad C_L := 1.3$ $\quad C_D := 0.025$

$z := 6$ $\quad C := 3.2\ in$ $\quad D_1 := 10\,in$

$D_2 := 15\,in$ $\quad r_2 := \frac{D_2}{2}$ $\quad \beta_2 := 22\ deg$

$\beta_1 := 45\,deg$ $\quad r_1 := \frac{D_1}{2}$

$$r_m^{\ 2} = \frac{(r_1^{\ 2} + r_1^{\ 2})}{2}$$

$$r_m := \sqrt{\frac{\left(r_1^{\,2} + r_2^{\,2}\right)}{2}} = 6.374\,in$$

$$s := \frac{2 \cdot \pi \cdot r_m}{z} = 6.675\ in$$

$$U_m : r_m \cdot \omega = 83.433\,\frac{ft}{s}$$

$$C_a := \frac{U_m}{\tan(\beta_1)} = 83.4333\,\frac{ft}{s}$$

$$Q := C_a \cdot \pi \cdot \left(r_2^{\,2} - r_1^{\,2}\right) = 56.882\ \frac{ft^3}{s}$$

$$\tan(\beta_m) = \frac{1}{2} \cdot \left(\tan(\beta_1) + \tan(\beta_2)\right)$$

$$\beta_m := \operatorname{atan}\left(\frac{1}{2} \cdot \left(\tan(\beta_1) + \tan(\beta_2)\right)\right) = 35.069\,deg$$

$$W_m := \frac{C_a}{\cos(\beta_m)} = 101.939\,\frac{ft}{s}$$

$$r_m := \sqrt{\frac{\left(r_1^{\,2} + r_2^{\,2}\right)}{2}} = 6.374\,in$$

$$s := \frac{2 \cdot \pi \cdot r_m}{z} = 6.675\,in$$

$$U_m := r_m \cdot \omega = 83.433\ \frac{ft}{s}$$

$$C_a := \frac{U_m}{\tan(\beta_1)} = 83.433\ \frac{ft}{s}$$

$$Q := C_a \cdot \pi \cdot \left(r_2^{\,2} - r_1^{\,2}\right) = 56.882\ \frac{ft^3}{s}$$

$$tan(\beta_m) = \frac{1}{2} \cdot \left(tan(\beta_1) + tan(\beta_2)\right)$$

$$\beta_m := \operatorname{atan}\left(\frac{1}{2} \cdot \left(tan(\beta_1) + tan(\beta_2)\right)\right) = 35.069\,deg$$

$$W_m := \frac{C_a}{\cos(\beta_m)} = 101.939\,\frac{ft}{s}$$

$$\rho{:}62.4\frac{lb}{ft^3} \qquad g := 32.2\frac{ft}{s^2}$$

$$\Delta p := \frac{1}{2}\cdot\frac{\rho}{g}\cdot W_m^{\ 2}\cdot\left(\frac{C}{s}\right)\cdot\left(\left(C_L\cdot\sin\left(\beta_m\right)\right)-\left(C_D\cdot\sin\left(\beta_m\right)\right)\right)$$

$$\Delta p := \frac{1}{2}\cdot 62.4\frac{lb}{ft^3}\cdot\left(103.313\frac{ft}{s}\right)^2\cdot\left(\frac{3.0\,in}{7.025\,in}\right)\cdot\left(1.32\cdot\sin\left(\left(\beta m\right)-0.027\cdot\cos\left(\beta_m\right)\right)\right)$$

$$= 22.542\ psi$$

4.13 EXERCISE PROBLEMS

Problem 4.1

An axial flow pump delivers water at a flow rate of 1.2 m^3/s and 12 m head while rotating at 600 rpm. The tangential velocity at the outlet is 20 m/s and the overall pump efficiency is 78%. Determine:

(a) the power delivered to the water
(b) the power input to the pump
(c) the impeller tip diameter if the hub diameter is 30 cm.

Problem 4.2

An axial flow pump discharges 50 l/s while running at 1100 rpm. The head generated is 5 m at the overall efficiency of 68%. If outer diameter is 40 cm and hub diameter is 20 cm, determine the flow velocity, assuming it to be constant from hub to tip. Also determine the power input to the pump.

Problem 4.3

An axial flow impeller rotates at 1500 rpm. The number of blades is six, inlet blade angle β_1 is 32°, and outlet blade angle β_2 is 18°. Neglecting the drag effect determine the pressure rise at the mean radius. Draw the velocity triangles. The chord length is 10 cm. D_1 = 20 cm and D_2 = 30 cm.

Problem 4.4

An axial flow pump impeller consists of six blades with chord length C = 3.2 in. Hub diameter D_1 = 8 in and tip diameter D_2 = 14 in. If C_L = 1.4 and C_D = 0.026 and the impeller is rotating at 1200 rpm, determine the flow rate and static pressure across the impeller. Assume blade angle $\beta_1 = 40°$ and $\beta_2 = 20°$.

Problem 4.5

An axial flow impeller rotates at 1200 rpm. If the number of blades is five and inlet blade angle is 28° and outlet blade angle is 15°, determine the pressure rise at the mean radius, neglecting the drag effect. Draw the velocity triangles. The chord length is 8 cm. $D_1 = 10$ cm and $D_2 = 25$ cm.

Problem 4.6

An axial flow impeller rotates at 1200 rpm. If the number of blades is six, inlet blade angle is 30°, and outlet blade angle is 15°, determine the pressure rise at the mean radius, neglecting the drag effect. Draw the velocity triangles. The chord length is 3.8 in. $D_1 = 4$ in and $D_2 = 10$ in.

4.14 BIBLIOGRAPHY

Blaha, J. and Brada, K. *Pumping Technology Manual*. Prague: ČVUT, 1997.

Brada, K. and Bláha, J. *Hydraulic Machines (in Czech)*. Prague: SNTL, 1992.

Cheng, L., Liu, C., Luo, C., Zhou, J. R. and Jin, Y. "Research on the unstable operating region of axial-flow and mixed flow pump," in *IOP Conference Series: Earth and Environmental Science*, 15 (2012) 032050. Beijing: IOP Publishing, 2012.

Dixon, S. L. and Hall, C. A. *Fluid Mechanics and Thermodynamics of Turbomachinery, 7th edition*. Oxford: Elsevier, 2013.

Fazil, J. and Jayakumar, V. "Investigation of airfoil profile design using reverse engineering Bezier curve," *J. Eng. Appl. Sci.*, 6 (7), 43, 2011.

Geerts, S. "Experimental and numerical study of an axial flow pump," Brussels: Vrije Universiteit Brussel, 2006.

Gülich, J. F. "Operation of centrifugal pumps," in *Centrifugal Pumps*. Prague: ČVUT, 2020.

Karassik, I. J., Krutzsch, W. C., Fraser, W. H. and Messina, J. P. *Pump Handbook*. New York: McGraw Hill, 1976.

Kothandaraman, R. and Rudramoorthy, C. P. *Fluid Mechanics and Machinery*, second edition. New Delhi: New Age International, 2007.

Lobanoff, V. S. and Ross, R. R. *Centrifugal Pumps: Design and Application*. Houston, TX: Gulf Professional Publishing, 2013.

Melichar, K. B. and Blaha, J. *Hydraulic Machines: Construction and Operation (in Czech)*. Prague: CVUT, 2002.

Mikhailov, A. K. and Maliushenko, V. V. *Lobe Pumps. Theory, Calculation and Design Engineering* (in Russian). Moscow: Mashinostroenie Publ., 1977.

Neumann, B. *The Interaction Between Geometry and Performance of a Centrifugal Pump*. London: Mechanical Engineering Publications, 1991.

Paciga, G. M. and Strýček, O. *Pumping Techniques-in Slovak*. Bratislava, Slovakia: ALFA, 1984.

Rama S. R. Gorla and Khan, A. A. *Turbomachinery Design and Theory*. New York: Marcel Dekker, 2003.

Steponoff, A. J. *Centrifugal and Axial Flow Pumps: Theory, Design and Application*, second edition. Malabar, FL: Krieger, 1957.

Stepanoff, A. J. *Pumps & Blowers & Two Phase Flow*. Malabar, FL: Krieger, 1966.

Strýček, O. *Hydrodynamic Pumps*. Bratislava, Slovakia: STU Bratislava, 1994.

Sultanian, B. K. *Logan's Turbomachinery*. Boca Raton, FL: CRC Press, 2019.

Turton, R. K. *Principles of Turbomachinery*. London: Chapman and Hall, 1984.

Varchola, P. H. M. and Bielik, T. "Methodology of 3D hydraulic design of a impeller of axial turbo machine," *Eng. Mech.*, 20, 107–118, 2013.

Varchola, M. and Hlobocan, P. "Hydraulic interaction between an impeller and axial diffuser of a mixed-flow pump-in Slovak," in *Current Trends in Development of Pumping Machinery*, 2013. Applied Mechanics and Materials, 630, 35–42, available at: www.scientific.net/AMM.630.35.

Varchola, M. and Hlbočan, P. *Hydraulic Design of an Axial Machine*. Bratislava, Slovakia: STU Bratislava, 2015.

Varchola, M. and Hlobocan, P. *Hydraulic Design of Centrifugal Pumps*. Bratislava, Slovakia: STU Bratislava, 2016.

Xu, L., Ji, D., Shi, W., Xu, B., Lu, W. and Lu, L. "Influence of inlet angle of guide vane on hydraulic performance of an axial flow pump based on CFD," *Shock Vib.*, 2020, 8880789.

Yang, K. F., Feng, J. J., Zhu, G. J., Lu, J. L. and Luo, X. Q. "Study on improvement of pump characteristic of an axial flow axial pump by grooving inlet wall," in *IOP Conference Series: Earth and Environmental Science*, 2018.

5 Hydro-Turbines

5.1 INTRODUCTION

The first mention of the use of hydropower dates to 1000 BC, when water wheels were used to pump water into irrigation canals (in India, China, Egypt). The most frequent use of hydropower was in mills and sawing lumber with water-powered sawmills. The development of hydropower plants (hydropower) was driven by the development of water turbines. Since the beginning of the 20th century, there has been a rapid development of hydroelectric power plants and thus of water turbines. Hydropower is the oldest and most established source of electricity generation. Its industrial use goes back to 1880 when a small direct-current power plant was built in Wisconsin, USA. Energy growth has since magnified. Hydropower now has a global annual growth rate of about 5% (i.e., doubling by volume every 15 years). Hydropower in 2006 had an installed capacity of 843 GW worldwide, of which 770 GW are large hydropower plants and 76 GW are small hydropower plants. Large hydropower plants account for 15% of world electricity production. The growth of hydropower can best be seen in Figure 5.1, which shows an increase in performance.

Hydropower in 2019 had an installed capacity of 1308 GW worldwide. Electricity generation from hydropower hit a record 4306 terawatt hours (TWh), the single greatest contribution from a renewable energy source in history. The largest waterworks in the world are listed in Table 5.1. Table 5.2 gives a classification of hydropower plants by power output. China is the largest producer of hydropower today, followed by Canada, Brazil, and the USA. Because of its abundance, hydropower is projected to be the dominant energy resource in the world for many years to come. Countries building large hydropower plants with capacity exceeding 5000 MW include China (Baihetan and Wudongde hydropower plant), Myanmar (TaSang hydropower plant), and Ethiopia.

5.2 HYDROPOWER PLANTS

5.2.1 WATER/HYDROLOGIC CYCLE

The natural water (aka hydraulic) cycle is a continuous movement by which water circulates between the earth's oceans, atmosphere, and land. This process is usually represented by a simplified circular cycle involving evaporation, condensation, and precipitation. The process is much complicated because the paths that water follows as it moves through the earth's ecosystem are complex and not fully understood.

DOI: 10.1201/9781003007142-5

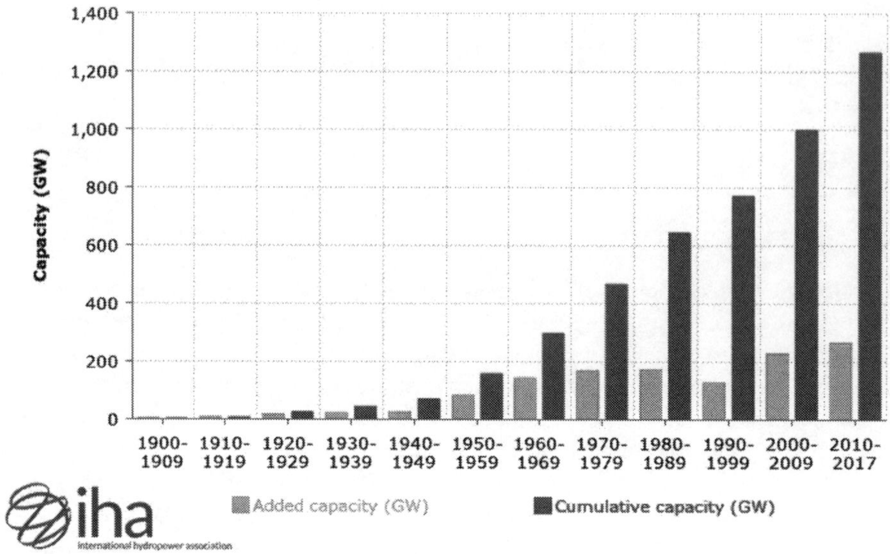

FIGURE 5.1 Hydropower growth since 1900 (source: www.hydropower.org/discover/history-of-hydropower).

TABLE 5.1
Classification of Hydropower Plants

Classification	Power output
Large	>100 MW
Medium	10–100 MW
Small	1–10 MW
Mini	100 kW–1 MW
Micro	5–100 kW
Pico	<5 kW

Source: Dilip Singh, Micro Hydro Power: Resource Assessment Handbook, APCTT, 2009.

In the water cycle liquid water evaporates into water vapor which condenses to form clouds, and finally precipitates back to earth in the form of rain and snow. The movement of the water is shown in Figure 5.2. After the water reaches the ground some may evaporate back into the atmosphere; some may penetrate the surface and become groundwater. The groundwater either soaks its way into the oceans, rivers, and streams, or is released back into the atmosphere through transpiration. The water remaining on the earth's surface is called runoff water, which discharges into streams, rivers, and lakes and finally is carried back to the ocean, starting the cycle again. Ice

TABLE 5.2
Largest Hydropower Plants in the World

Name	Country	Year of Completion	Total Capacity (MW)	Max. Annual Electricity Production (TW-hour)	Area Flooded (km^2)
Three Gorges Dam	China	2012	22,500	>100	1045
Itaipu	Brazil and Paraguay	1986/1991/2003	14,000	90	1350
Guri (Simón Bolívar)	Venezuela	1986	10,200	46	4250

FIGURE 5.2 The water cycle (www.noaa.gov/).

and snow can change directly into vapor by a process known as sublimation. When water vapor converts to solid it is called deposition.

Considering the water cycle, the energy of water can be considered an inexhaustible, constantly renewing source of energy. This is advantageous when compared to some renewable energy sources, such as solar and wind, which are intermittent. The intensity of the wind is variable, becoming zero during windless times. The intensity of solar radiation also depends on conditions such as clouds, the season, and time of day. From this point of view, the energy of the water flow has the advantage that it can be used continuously with a suitably chosen design solution of the water turbine during the entire life of the turbine.

In hydropower plants the kinetic energy of water is converted into mechanical work using water/hydro-turbines. A water turbine is a rotating mechanical machine which is connected to an electric generator through a shaft to generate electricity. The water turbine together with the electric generator forms the fundamental part of hydroelectric power plants. Hydropower is considered as a renewable source of energy because water supply is constantly replenished by the water cycle. Hydropower is a never-ending resource and eco-friendly because it does not emit greenhouse gases to the atmosphere that contribute greatly to pollution and global warming.

5.2.2 Classification of Hydropower Plants

Hydropower plants can be classified into three categories: impoundment, diversion, and pumped storage.

5.2.2.1 Impoundment Hydropower Plants

Impoundment hydropower plants are the most common type of hydroelectric power plants and consist of a dam to store river water in a reservoir (Figure 5.3). When electricity is needed, water from the reservoir flows through a penstock and hits a turbine, rotating it. This in turn spins an electrical generator which is connected to the turbine through a shaft to produce electricity. One of the world's most famous impoundment dams is the Hoover Dam.

5.2.2.2 Diversion Hydropower Plants

Diversion (aka run-of-river or ROR) hydropower plants (Figure 5.4) channel flowing water from a river through a penstock or canal to rotate a turbine. Diversion hydropower plants do not have water storage dams. Diversion hydropower plants supply electricity, with some flexibility of operation for daily fluctuations in demand through water flow that is regulated by the facility.

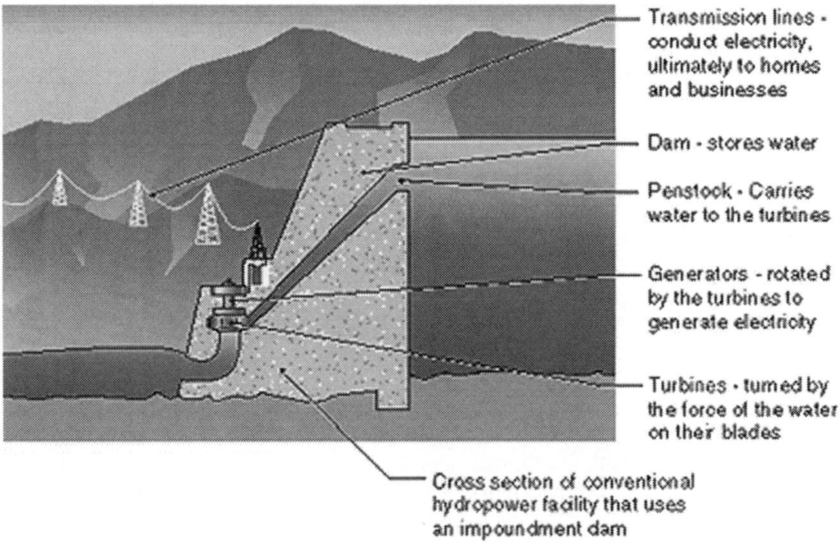

FIGURE 5.3 Schematic of an impoundment hydropower plant (www.energy.gov/).

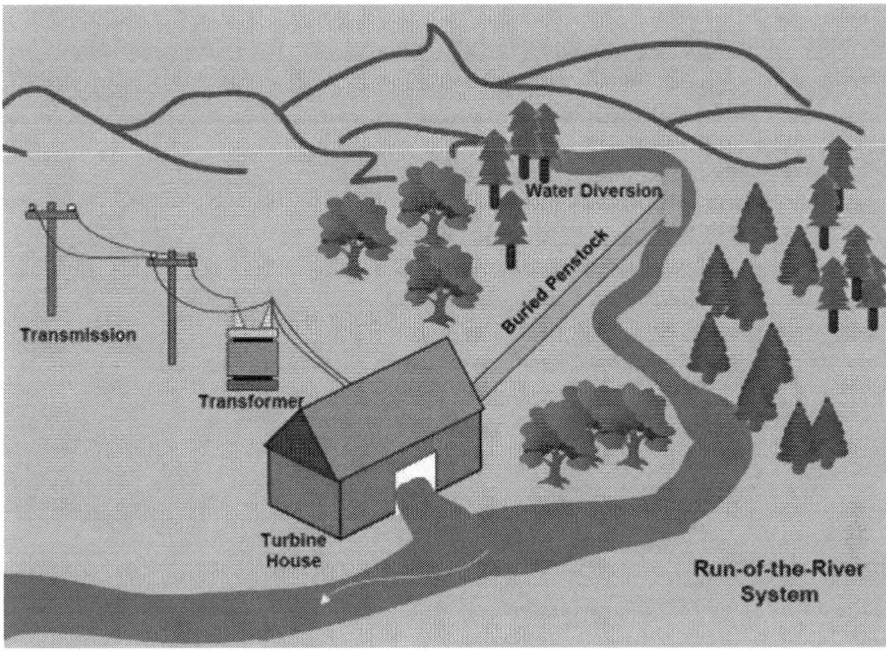

FIGURE 5.4 Schematic of a run-of-river (ROR) hydropower plant (www.unidata.com.au/).

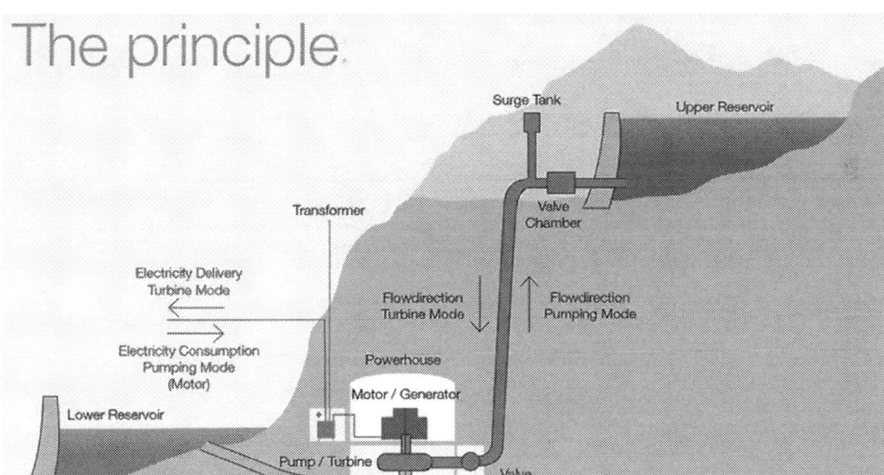

FIGURE 5.5 Schematic pumped storage hydropower plant (www.andritz.com/).

5.2.2.3 Pumped Storage Hydropower Plants

Pumped hydropower plants (Figure 5.5) store water in an upper reservoir, pumped from a reservoir at a lower elevation. During peak power demand periods, electricity is generated by releasing the stored water to spin the turbines like an impoundment hydropower plant. During low power demand (usually at night or on the weekend when electricity price is also at lower cost), energy is stored by pumping water from the lower reservoir to the upper reservoir. Lower-cost electricity from the grid to pump the water back to the upper reservoir is utilized. In pumped storage power plants, reversible pump–turbine/motor–generator assemblies are used as both pumps and turbines. Pumped storage hydropower plants are unlike traditional hydropower plants in that they are a net consumer of electricity, due to hydraulic and electrical losses incurred in the cycle of pumping from lower to upper reservoirs. The round-trip efficiencies of these hydropower plants can exceed 80% and are very economical due to peak and off-peak price differentials and their potential to provide critical auxiliary grid services.

5.3 HYDRO-TURBINES

5.3.1 Brief History of Hydro-Turbines

Hydro-turbines have had a long period of development. The oldest and simplest form is the water wheel, which was first used in ancient Greece and then used to grind grain throughout medieval Europe. Hydro-turbines as we now know them, based on the theoretical foundations developed by Leonard Euler, were built at the beginning of the 19th century by a French engineer. Benoît Fourneyron developed the first commercially successful water turbine in 1827 in St Blasien. Later, Fourneyron designed and built water turbines for industrial use, which achieved outputs of about 50 hp at an efficiency of more than 80% and a speed of 2300 rpm. The development of hydro-turbines is listed chronologically as follows.

- 1750 J. A. Segner – reaction turbine
- 1750–1754 Euler develops the turbine theory
- 1827 Forneyron – radial turbine, 30–40 hp, $D = 0.5$ m, $n = 2300\ 1 / \text{min}$
- 1840 Henschel / Jonval – axial turbine
- 1840 Jonval – draft tube
- 1849 Francis – Francis turbine; Fink – load regulation with guide vanes
- 1890 Pelton – Pelton turbine, impulse turbine
- 1913 Kaplan – Kaplan turbine, propeller
- 1955 Deriaz – Deriaz turbine.

5.3.2 Classification of Hydro-Turbines

The two main types of hydro-turbines are impulse and reaction. Generally, reaction turbines are used for a water source with low head and high flow rate. Impulse turbines are used for a water source with high head and low flow rate. This means the type of hydro-turbine selection depends on the height of standing water, which is called "head," and the flow or volume of water available at the site. Other deciding factors include how deep the turbine must be set, efficiency, and cost. Figure 5.6 gives a classification of hydro-turbines. The most widely used hydro-turbines will be briefly discussed below.

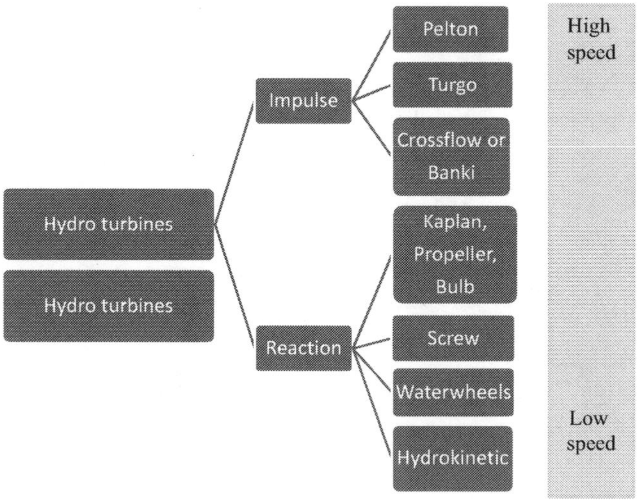

FIGURE 5.6 Classification of hydro-turbines.

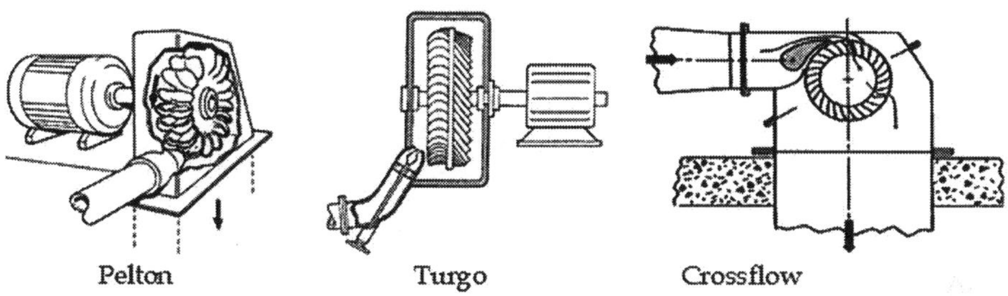

FIGURE 5.7 Types of impulse turbine (Source: Paish, 2002).

5.3.3 Impulse Hydro-Turbines

Impulse turbines use the velocity of the water to move the runner. The water jet hits each bucket on the runner (Figure 5.7). There is no suction on the downside of the turbine, and the water flows out of the bottom of the turbine housing after hitting the runner. Impulse turbines are generally suitable for high-head, low-flow applications.

5.3.3.1 Pelton Turbines

The Pelton wheel is an impulse-type hydro-turbine and has one or more free jets impinging on the buckets of a runner. Figure 5.8 shows operation of a Pelton turbine. In Figure 5.8, the penstock pipe delivers the water on the left-hand side. Before entering the turbine the water passes through the nozzle and then impinges on the bucket. The spear jet adjusts the flow rate through the turbine by moving back and forth, thus varying the flow area and hence the flow rate.

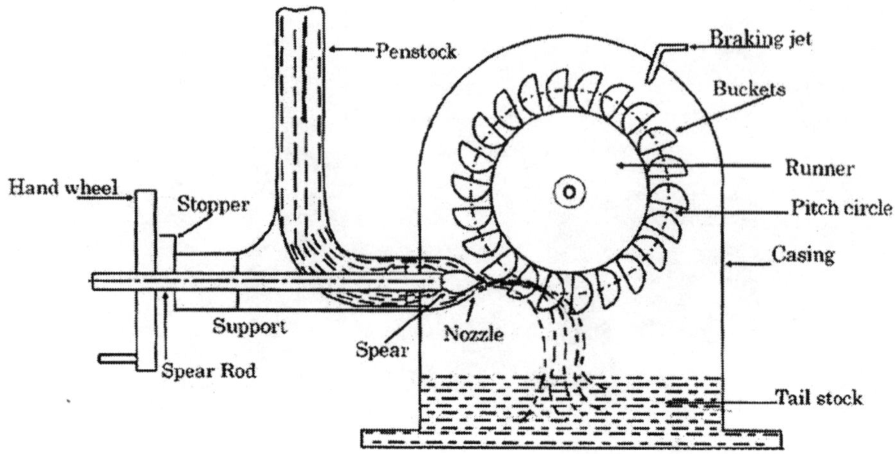

FIGURE 5.8 A Pelton wheel.

5.3.3.2 Turgo Turbines

A Turgo turbine is a variation on the Pelton and is manufactured by Gilkes in England. The Turgo turbine can handle higher flow rate than a Pelton turbine of physically the same size and the rotor is slightly cheaper to manufacture. The main physical difference between a Pelton and Turgo wheel is in the design of the bucket. With Pelton turbines each bucket consists of two symmetrical halves with a shape of semi-ellipsoidal cup discharging the water outward from the side of the buckets. These symmetrical parts are divided by a sharp-edged ridge called a splitter. Water, coming out of one or more nozzles at high speed, hits the buckets to convert all its available energy into kinetic energy of the runner. In a Turgo turbine the water jet strikes one side of the rotor and exits from the opposite side (Figure 5.9).

5.3.3.3 Crossflow (Bánki) Turbines

A crossflow turbine, also known as Bánki–Michell or Ossberger turbine, is another impulse-type turbine (Figure 5.10). It looks like a "squirrel cage" blower. The water enters as a flat sheet rather than a round jet. It is guided on to the blades, travels across the turbine, and hits the blades, allowing the water to flow through the blades twice. A guide vane is used at the entrance to direct the flow to a limited portion of the runner. The crossflow is useful for larger water flows and lower heads than the Pelton.

5.3.4 Reaction Hydro-Turbines

Reaction turbines fill a volute with whirling water that rotates the runner blades. In a reaction turbine, the impeller is enclosed in a casing and therefore the water is always at a pressure other than atmosphere. In reaction turbines, the water pressure changes as it moves through the turbine blades. The runner of a reaction turbine needs to be submersed in a water stream flowing over the blades instead of striking

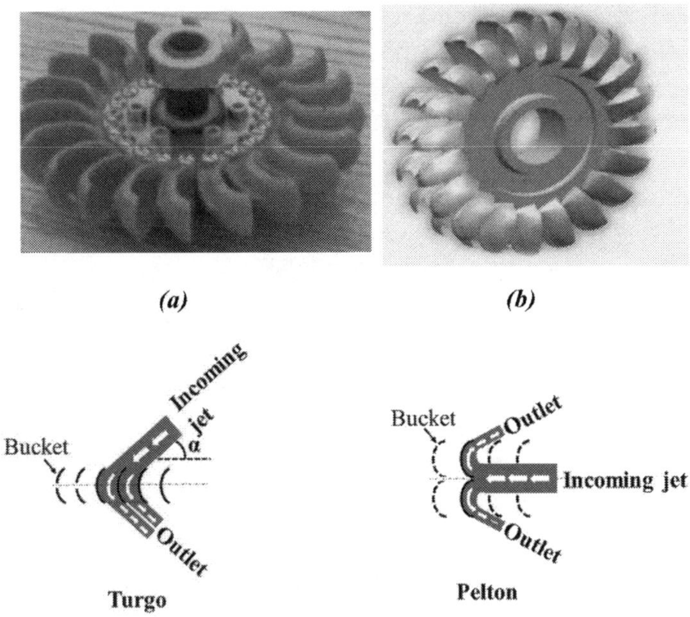

FIGURE 5.9 Turgo (a) and Pelton turbine (b) runners.

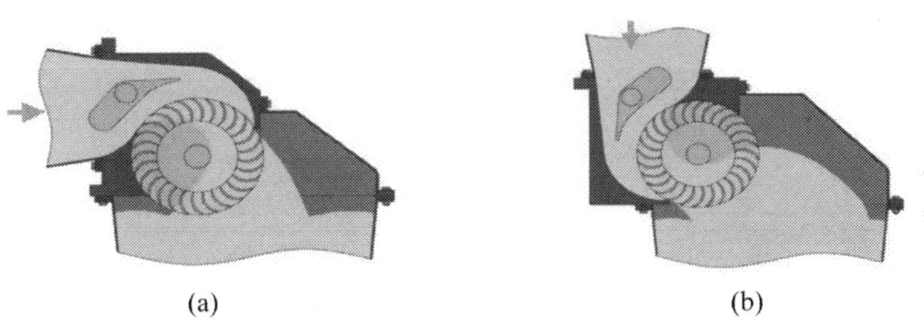

FIGURE 5.10 A crossflow turbine (a) with a horizontal entry and (b) with a vertical entry (www.ossbergerhydro.com/).

each individually. Reaction turbines develop power from the combined action of pressure and moving water. Thus, water leaving the blade has a large relative velocity but small absolute velocity. Therefore, most of the initial energy of water is given to the impeller. In reaction turbines, water leaves the impeller at atmospheric pressure. The pressure difference between entrance and exit points of the impeller is known as reaction pressure. Reaction turbines are used for sites with lower head and higher flows compared to impulse turbines which require high head and low flow rate. Types of reaction turbines include Francis, Kaplan, propeller, bulb, kinetic, and Straflo turbines.

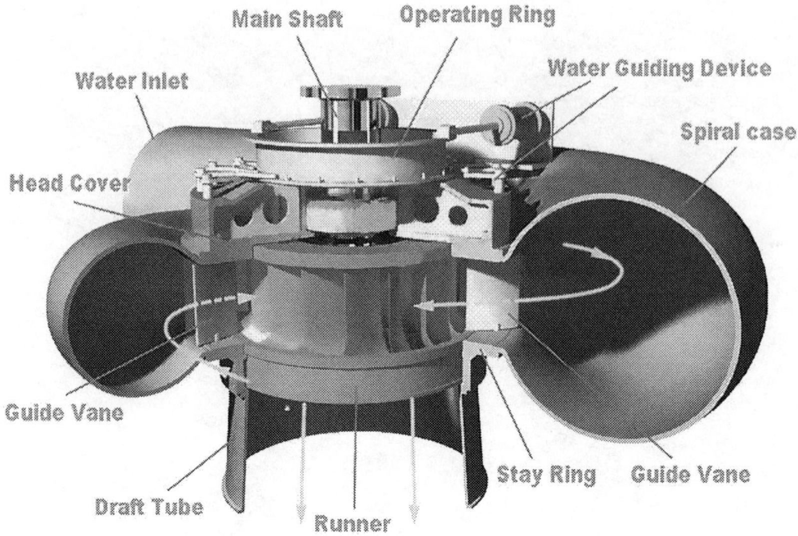

FIGURE 5.11 Francis reaction turbine (www.mechanicalbooster.com).

5.3.4.1 Francis Turbines

The radial flow or Francis turbine is a reaction machine. As the water flows over the curved blades, the pressure head is transformed into a velocity head. Developed by an American engineer, James Bichens Francis, Francis turbines are an inward-flow reaction turbine that combines radial and axial flow concepts (Figure 5.11).

In Francis turbines water changes direction as it passes through the turbine, hence the alternative name mixed-flow turbine. The water enters the turbine in a radial direction, flowing towards its axis, strikes and interacts with the turbine blades, and exits along the direction of that axis.

5.3.4.2 Kaplan Turbines

Developed in 1913, by an Austrian professor Viktor Kaplan, the Kaplan turbine (Figure 5.12) can work at low head and high flow rates very efficiently. In a Kaplan turbine water flows through the runner along the direction parallel to the axis of rotation of the runner, which means that the flow direction does not change as it crosses the rotor. A Kaplan turbine has adjustable blades inside a tube. Flow rate is regulated by opening and closing the guide vanes. When the guide vanes are fully closed, they will stop the water completely and bring the turbine to rest.

5.3.4.3 Propeller Turbines

Propeller turbines, like Kaplan turbines, are axial flow machines in which the flow through the runner is predominantly axial. The difference between propeller and Kaplan turbines is that the propeller turbine has fixed runner blades while the Kaplan turbine has adjustable runner blades.

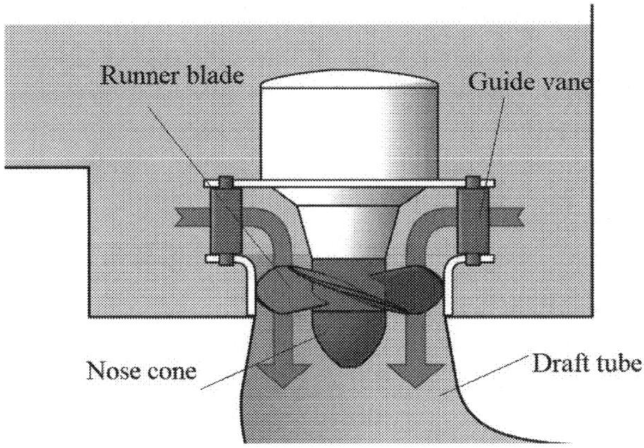

FIGURE 5.12 Kaplan reaction turbine (www.mechanicalbooster.com).

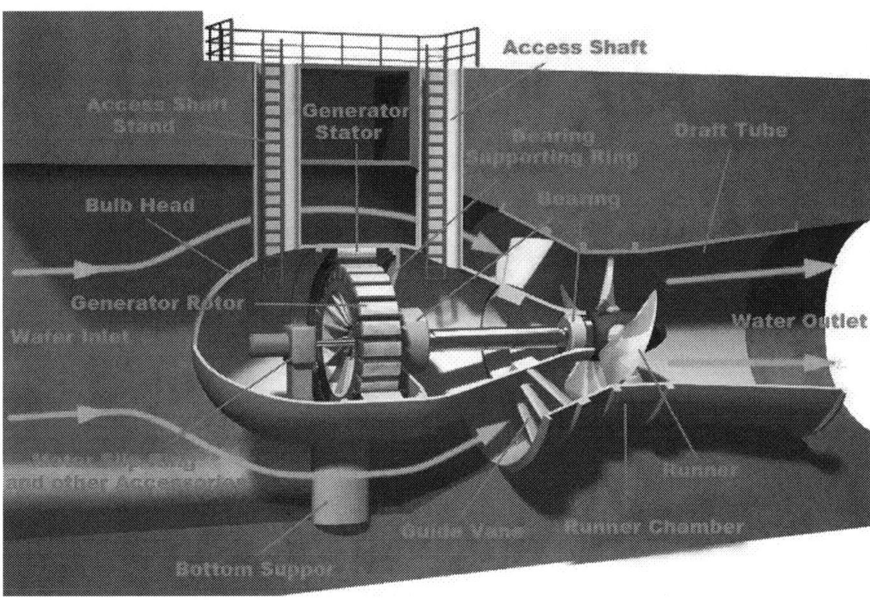

FIGURE 5.13 Bulb reaction turbine (www.eternoohydro.com).

5.3.4.4 Bulb Turbines

Bulb turbines (aka tubular turbines) are reaction turbines which can be useful for exploitation of tidal and hydraulic power with extremely low water head and extremely large discharge. The turbine components as well as the generator are housed inside a "bulb" (Figure 5.13). The water flow direction in bulb turbines is mixed axial–radial into the guide vane cascade. Bulb turbines can be utilized to tap electrical power from fast-flowing rivers on the hills.

5.3.5 Euler Turbine Equations

The fluid velocity at the inlet and outlet of the turbine runner can have three components: tangential, axial, and radial. This means that the momentum of a fluid also has three components at the inlet and outlet. This also means that the force acting on the runner can have three components. Of these, only the tangential force component causes the runner to rotate and perform the work. The axial component creates thrust in the axial direction and is parallel to the axis of rotation, which is captured by suitable axial bearings. The radial component passing through the axis of rotation cannot generate torque but creates a bending moment, which is again captured by the bearings.

The circumferential component of velocity is important to determine the work done and then the power. The force in the circumferential direction is equal to the product of the mass flow rate and the circumferential component of the absolute velocity. Because the tangential component of velocity changes with the radius, the work will also change with the radius. Applying conservation of angular momentum, we note that the torque, M_k, must be equal to the time rate of change of angular momentum of the water stream that flows through the device.

$$M_k = \rho Q c_{u1} r_1 - \rho Q c_{u2} r_2 = \rho Q \left(c_{u1} r_1 - c_{u2} r_2 \right) \tag{5.1}$$

where the torque at the inlet is given by $\rho Q c_{u1} r_1$ and the torque at the outlet is given by $\rho Q c_{u2} r_2$. The work per unit time, or power, P, is the torque multiplied by the angular velocity,

$$P = M_k \varpi = M_k = \rho Q c_{u1} r \varpi_1 - \rho Q c_{u2} r_2 \varpi = \rho Q \left(c_{u1} u_1 - c_{u2} u_2 \right) \tag{5.2}$$

Since power, P, can also be written as $P = \rho Q g H$, then

$$P = \rho Q g H = \rho Q \left(c_{u1} u_1 - c_{u2} u_2 \right) \rightarrow gH = c_{u1} u_1 - c_{u2} u_2 \tag{5.3}$$

Hence the Euler turbine equation in head form becomes

$$H = \left(\frac{c_{u1} u_1 - c_{u2} u_2}{g} \right) \tag{5.4}$$

where

c_1, c_2 are absolute velocities at the inlet and outlet respectively
u_1, u_2 are tangential velocity components at the inlet and outlet
c_{u1}, c_{u2} are tangential components of the absolute velocities at the inlet and outlet respectively
c_{m1}, c_{m2} are meridional components of the absolute velocity at the inlet and outlet
w_1, w_2 are relative velocities at the inlet and outlet
w_{u1}, w_{u2} are tangential components of the relative velocity at the inlet and outlet
Note that $w_{m1} = c_{m1}$, $w_{m2} = c_{m2}$
u_1, u_2 are tangential velocity components at the inlet and outlet.

The following equations can be derived from the velocity triangles of Figure 5.14.

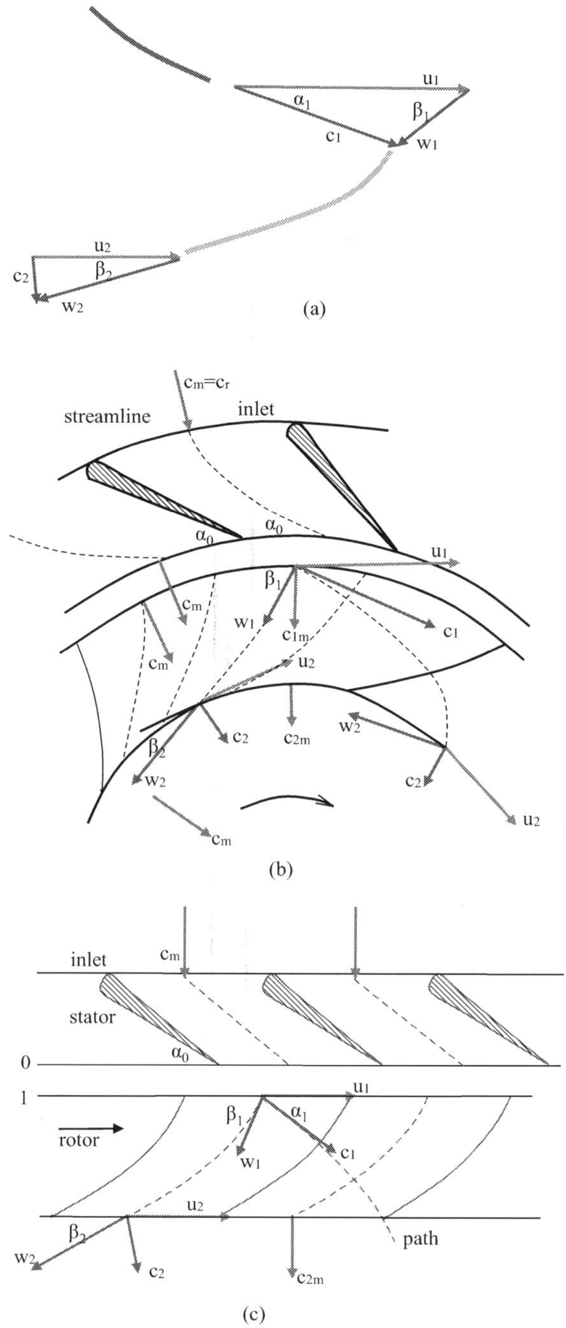

FIGURE 5.14 Velocity triangles: (a) at the inlet and outlet of a turbine blade; (b) shown with inlet guide vanes and the runner; (c) in cascade view.

$$w_1^2 = u_1^2 + c_1^2 - 2u_1 c_1 cos\,\alpha_1 = u_1^2 + c_1^2 - 2u_1 c_{u1} \tag{5.5}$$

Similarly, at the outlet

$$w_2^2 = u_2^2 + c_2^2 - 2u_2 c_2 cos\,\alpha_2 = u_2^2 + c_2^2 - 2u_2 c_{u2} \tag{5.6}$$

Substituting into Equation 5.3, we get

$$gH = \left(c_{u1}u_1 - c_{u2}u_2\right) = \frac{c_1^2 - c_2^2}{2} + \frac{u_1^2 - u_2^2}{2} - \frac{w_1^2 - w_2^2}{2} \tag{5.7}$$

In Equation 5.7, the components are

$\frac{c_1^2 - c_2^2}{2}$ dynamic component of the work done

$\frac{u_1^2 - u_2^2}{2}$ centrifugl component of work, present only in radial machines

$\frac{w_1^2 - w_2^2}{2}$ accelerating component of work done, present only in reaction turbines.

5.3.6 Turbine Similarity

There are several variables affecting the characteristics of hydraulic turbines and it is cumbersome to carry out experiments to find out the influence of each parameter separately. Hence dimensional analysis is used where parameters are grouped to minimize the number of experiments. The dimensional groups that have been found to be useful in efficiently assessing the performance of hydraulic turbines are given below.

The head coefficient or specific head

$$\Psi = \frac{H}{u^2 g} = \frac{gH}{n^2 D^2} \tag{5.8}$$

Flow coefficient or specific capacity

$$\varphi = \frac{Q}{nD^3} \tag{5.9}$$

Power coefficient or specific power

$$C_p = \varphi\Psi = \frac{P}{\rho n^3 D^5} \tag{5.10}$$

Dimensionless specific speed

$$n_s = \frac{nP^{\frac{1}{2}}}{\sqrt{\rho}(gH)^{\frac{5}{4}}} = \frac{n\sqrt{Q}}{(gH)^{\frac{3}{4}}}$$

where

P is power
g is acceleration due to gravity
H is head
n is rotational speed
Q is flow rate
ρ is density.

Dimensional specific speed

$$n_s = \frac{nP^{\frac{1}{2}}}{(H)^{\frac{5}{4}}} \tag{5.11}$$

In Equation (5.11), P is power in kW, n is in rpm, and H is in meter.

These dimensionless numbers are used in testing models and developing prototypes. Specific speed can be used to choose the type of turbine that provides the best efficiencies under given circumstances. Specific speed and efficiencies of the turbines are shown in Figure 5.15.

Generally scaling laws within families of similar machines are applied to examine the effect of a change in one or more operating variables, such as the effect of speed. The turbine similarity is given by the relationships between the values of head, flow rate, rotational speed, power, and diameter (H, Q, n, P, D). In the same turbine, flow Q changes with a change in head. For the same turbine, the change in flow rate results in a change in water velocity or rotational speed. This happens at constant blade angle. These facts indicate that similitude is preserved.

5.3.6.1 Specific Speed

The turbine specific speed is the speed of a geometrically similar turbine, which develops unit power under a unit head of water. Table 5.3 gives the operating ranges of commonly used turbines.

Specific speed is important in that it indicates the speed of a turbine which is identical in shape, blade angles, gate opening, etc. (i.e., a model) with the actual turbine (prototype). Specific speed indicates the flow area and shape of the runner. Generally, turbines with low specific speeds are suitable for low flow rate and high heads; and turbines with high specific speed are suitable for high flow rate and low heads. Operating ranges of different turbines as a function of flow rates and available

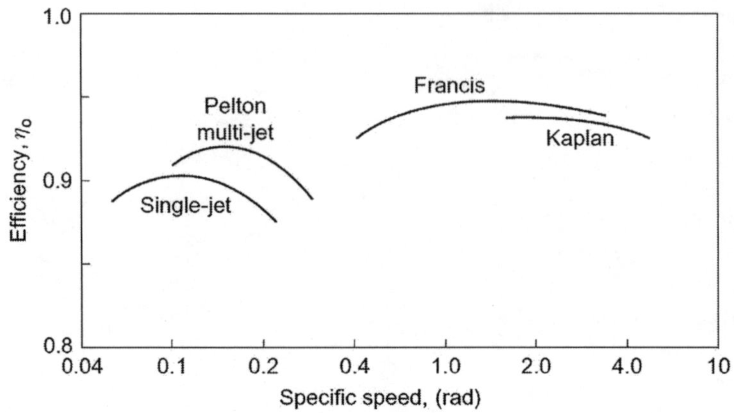

FIGURE 5.15 Variation of efficiency with specific speed (Dixon and Hall 2013).

TABLE 5.3
Operating Ranges of Hydraulic Turbines

	Pelton turbine	Francis turbine	Kaplan turbine
Specific speed (rad)	0.05–0.4	0.4–2.2	1.8–5.0
Head (m)	100–170	20–900	6–70
Maximum power (MW)	500	800	300
Optimum efficiency (%)	90	95	94

Note: Values shown in the table are only a rough guide and are subject to change.
Source: Dixon and Hall 2013.

head can also be read from Figures 5.16 and 5.17. These two figures can help in the preliminary selection of turbines.

5.3.6.2 Model and Prototypes

In design of hydraulic turbines, the practice is that test results on smaller model turbines are used as performance indicators for larger units. This helps avoid direct manufacturing and testing of larger hydraulic turbines. Testing is done on much smaller similar model turbines from which performance prediction of larger turbine units is made. To apply this principle of geometric similarity, kinematic and dynamic similarity of the model and prototype should be fulfilled (Chapter 2). If geometric, kinematic, and dynamic similarity hold, then specific speeds, head coefficients, and flow and power coefficient will be identical between a model and a larger turbine called a prototype. This enables prediction of performances of larger turbines from

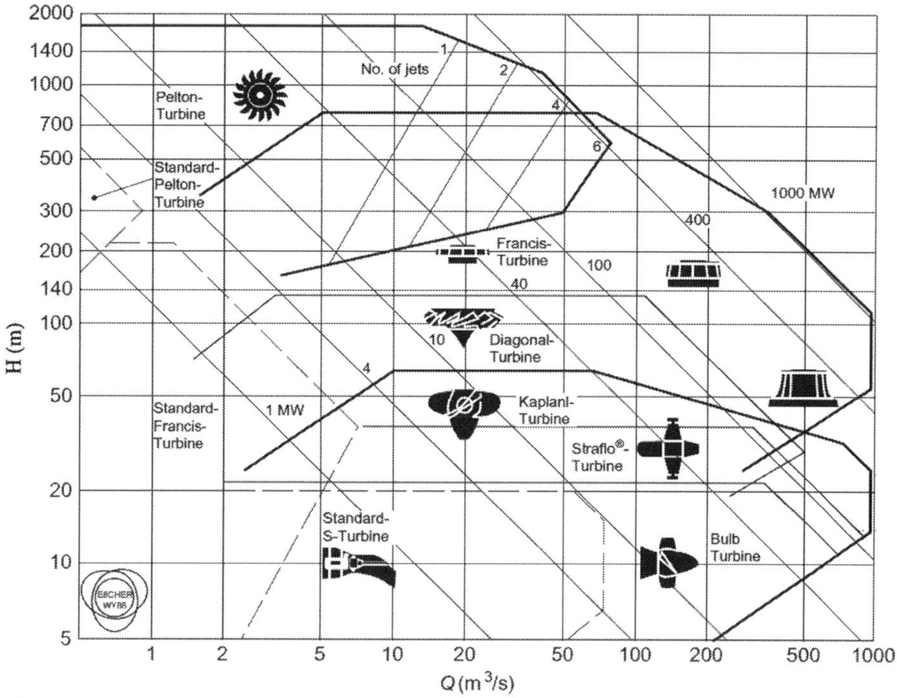

FIGURE 5.16 Application ranges of different types of hydraulic turbines (Dixon and Hall 2013).

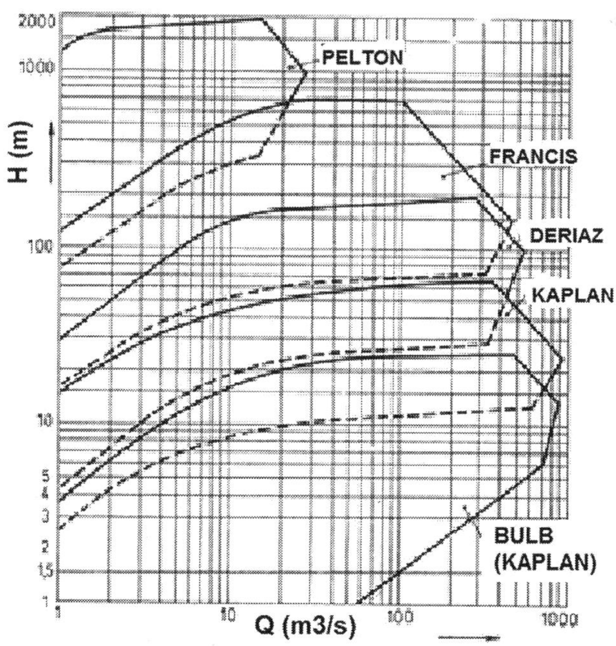

FIGURE 5.17 Turbine application depending on flow rate and head.

experimental results of model (smaller) turbines. Prediction can be made for operation in different head, speed, and flow conditions.

Reduced-scale physical models of turbines are used to determine the hydraulic behavior of prototypes. Most of the hydraulic parameters obtained in model tests are "similar" to prototype parameters according to similarity theories if they satisfy similitude requirements (geometric, kinematic, and dynamic). The relationship between model and prototype can be described using nondimensional coefficients as follows:

Using the head coefficient or specific head

$$\frac{H_m}{n_m^2 D_m^2} = \frac{H_p}{n_p^2 D_p^2} \tag{5.12}$$

Flow coefficient or specific capacity

$$\frac{Q_m}{n_m D_m^3} = \frac{Q_p}{n_p D_p^3} \tag{5.13}$$

Power coefficient or specific power

$$\frac{P_m}{n_m^3 D_m^5} = \frac{P_p}{n_p^3 D_p^5} \tag{5.14}$$

The above relationships can be used to determine the performance of a prototype based on experiments performed on models. This implies that the efficiencies of the model and the prototype are the same. This condition will not apply, especially when the differences in dimensions are very large (model and prototype). The efficiency of the turbine will depend on its dimensions so that larger machines have higher efficiency. The reason for discrepancies in model and prototype efficiencies is believed to be the difference in losses in the mode and prototype suggesting departures from strict dynamic similarity. Since exact mathematical treatment is not possible, empirical formulations from experimental data are presented. Two of these relations used for the conversion of efficiency from the model to the prototype are presented below.

$$\eta_p = 1-(1-\eta_m)\sqrt[4]{\frac{D_m}{D_p}}\sqrt[10]{\frac{H_m}{H_p}} \quad \text{Moody} \tag{5.15}$$

$$\eta_p = \frac{1}{1-\left(\frac{1}{\eta_m}--1\right)\left(0.3+0.7\sqrt[5]{\frac{Re_m}{Re}}\right)\left(Re=\frac{D\sqrt{2gH}}{\nu}\right)} \quad \text{Hutton} \tag{5.16}$$

5.3.6.3 Unit Quantities

Performance of a single turbine under different conditions can be predicted with the dimensionless coefficients described above. Consider a single turbine for which the head is changed from H to H'. When the turbine was operating under head H, let's say the absolute, the tangential, and relative velocities are c_1, u_1, and w_1. When the turbine is operating under head H', Q', and n', the velocity components become c'_1, u'_1, and w'_1. Then we can write:

$$\frac{H}{n^2 D^2} = \frac{H'}{n'^2 D^2} \rightarrow \frac{n'^2}{n^2} = \frac{H'}{H} \rightarrow \frac{n'}{n} = \sqrt{\frac{H'}{H}} \tag{5.17}$$

It is also possible to write:

$$n' = \frac{n}{\sqrt{H}} \quad \text{if } H' = 1\,\text{m} \tag{5.18}$$

Similarly, from the flow coefficient it is possible to get the following relationships:

$$\frac{Q}{nD^3} = \frac{Q'}{n'D^3} \rightarrow \frac{n'}{n} = \frac{Q'}{Q} \tag{5.19}$$

$$\frac{n'}{n} = \frac{Q'}{Q} = \sqrt{\frac{H'}{H}} \rightarrow Q' = \frac{Q}{\sqrt{H}} \quad \text{if } H' = 1\,\text{m} \tag{5.20}$$

Finally, from the power coefficient it is possible to get the following relationships:

$$\frac{P}{n^3 D^5} = \frac{P'}{n'^3 D'^5} \rightarrow \frac{n'^3}{n^3} = \frac{P'}{P} \rightarrow \frac{P'}{P} = \left(\frac{H'}{H}\right)^{\frac{3}{2}} \tag{5.21}$$

$$P' = \frac{P}{\sqrt{H^3}} \quad \text{if } H' = 1\,\text{m} \tag{5.22}$$

5.3.6.4 Diameter

Geometrically similar turbines of different diameters D, which operate under the same conditions, will have the same tangential and meridional velocities, which can be easily verified using Euler's turbine equation (Equation 5.4). For two turbines with diameters D_1 and D'_1, tangential velocity components can be written as:

$$u_1 = \frac{\pi n}{60} D_1 = \frac{\pi n'}{60} D'_1 \Rightarrow \frac{n}{n'} = \frac{D'_1}{D_1}$$

The flow rate Q depends only on the flow area. Therefore, the flow rate Q is given by:

$$Q = \frac{\pi D_1^2}{4} c_{m1} \text{ and } Q' = \frac{\pi D_1'^2}{4} c_{m1}, \text{ where } c_{m1} \text{ is meridional velocity}$$

$$\frac{D_1}{D'_1} = \sqrt{\frac{Q}{Q'}} = \frac{n'}{n} \tag{5.23}$$

The performance relationship is given by:

$$\frac{D_1}{D'_1} = \sqrt{\frac{QH}{Q'H}} = \sqrt{\frac{P}{P'}} = \sqrt{\frac{Q}{Q'}} = \frac{n'}{n} \tag{5.24}$$

5.3.6.5 Other Forms of Specific Speed

In addition to nondimensional specific speed, it is common to see dimensional specific speeds. As we have seen, the values of Q, n, D, and H of geometrically similar turbines are bound together by exact relations. Such turbines must have a common relationship, which we get from dimensional analysis. One such generally established relationship is specific speed. The specific speed of a turbine is defined as the speed of a geometrically similar turbine which would produce unit horsepower under unit head in meters and is defined below. From Equation 5.24, we get:

$$\frac{n_s}{n'} = \sqrt{\frac{P'}{P}} = \sqrt{\frac{P'}{1HP}} \Rightarrow n_s = n'\sqrt{\frac{P'}{0.736}} = 1.166 n'\sqrt{P'} \tag{5.25}$$

where n_s is specific speed. This equation is dimensionally nonhomogeneous.

Where n is in rpm, P is in kW.

$$n' = \frac{n}{\sqrt{H}} \text{ for H}' = 1 \text{ m and } P' = \frac{P}{\sqrt{H^3}} \text{ for H}' = 1 \text{ m}$$

$$n_s = n'\sqrt{P'} = 1.166\frac{n}{\sqrt{H}}\sqrt{\frac{P}{\sqrt{H^3}}} = 1.166\frac{n}{H}\sqrt{\frac{P}{\sqrt{H}}} \tag{5.26}$$

If we substitute into for power, $P = \rho QgH$, we get

$$n_s = 3.65n\sqrt{\frac{Q}{H^{3/2}}} = 3.65n\frac{\sqrt{Q}}{H^{0.75}} \tag{5.27}$$

Table 5.4 gives an approximate range of specific speeds for different types of turbines. Table 5.4 with Equations 5.25–5.27 can also be used as a guide in selecting appropriate type of turbine.

TABLE 5.4
Best Specific Speed Range for Different Types of Hydraulic Turbines

n_s	Type of Turbine
4–35	One-jet Pelton turbine
17–50	Two-jet Pelton turbine
24–70	Four-jet Pelton turbine
80–120	Slow-running Francis turbine
120–220	Normal Francis turbine
220–350	High-speed Francis turbine
350–450	Express Francis turbine
300–1200	Axial flow Kaplan turbine

5.3.7 Turbine Efficiencies

The gross head available at the hydropower plant site depends on the topographical conditions of the location. The gross head, H_g, is defined as the elevation difference between the upper and lower reservoirs. Head losses, h_f, occur during water transportation from the site to the turbine. The hydropower at the site is given by: $P = \rho gQH$. This power delivered by the turbine is less than this hydropower due to losses. This head is called the net head, H_{net} (Figure 5.18). In the following we will discuss different types of efficiencies, which account for these losses.

5.3.7.1 Hydraulic Efficiency

When analyzing the hydraulic efficiency of the turbine, we consider the available head to the turbine (Figure 5.18). The turbine does not completely convert the available waterpower to it, for example due to the friction between the fluid and the rotor. These losses are accounted for by hydraulic efficiency of the turbine. It is the ratio of power produced by the runner of the turbine to the waterpower supplied to it. In general, we can express the hydraulic efficiency by η_h.

$$\eta_h = \frac{Power\ produced\ by\ the\ runner}{\rho QgH} \tag{5.28}$$

5.3.7.2 Volumetric Efficiency

It is possible some water flows out through the clearance between the impeller and casing without passing through the impeller. Volumetric efficiency is defined as the ratio between the volume of water flowing through the impeller and the total volume of water supplied to the turbine. We express the volumetric efficiency by η_v.

$$\eta_v = \frac{Q - \Delta Q}{Q} \tag{5.29}$$

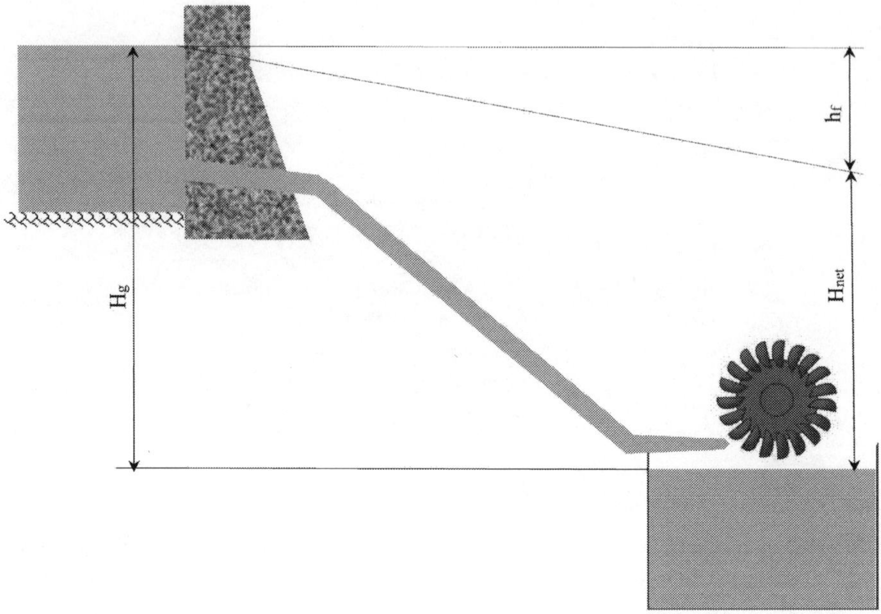

FIGURE 5.18 Gross and net head of a turbine.

5.3.7.3 Mechanical Efficiency

The power produced by the impeller is always greater than the power available at the turbine shaft. This is due to mechanical losses at the bearings, windage losses, and other frictional losses. The mechanical efficiency is given by η_m

$$\eta_m = \frac{Power\ available\ at\ the\ turbine\ shaft}{Power\ produced\ by\ the\ runner} \tag{5.30}$$

5.3.7.4 Overall Efficiency

This is the ratio of power output at the shaft and power input by head water at the turbine inlet.

$$\eta_t = \frac{Power\ available\ at\ the\ turbine\ shaft}{\rho Q g H} \tag{5.31}$$

$$\eta_t = \eta_h \eta_v \eta_m \tag{5.32}$$

5.4 PELTON TURBINE

5.4.1 Fundamental Theory

Pelton turbines work on impulse principle, in which high-speed jet water discharged from a nozzle (stator) strikes buckets of a turbine, inducing an impulsive force. Hence the names impulse turbine and impulse wheel. A Pelton turbine is particularly suitable for high head and low flow rate. The rotor of a Pelton turbine consists of a circular disc with several curved blades in the shape of a spoon (usually known as "buckets"). A high-speed water jet is directed towards the centers of these buckets using one or more nozzles (Figure 5.19). Therefore, Pelton turbines use the velocity of the water jet to rotate the runner by hitting the buckets. This strike of water jet causes a change in momentum, resulting in a force, which is applied to the buckets. At the center of the buckets is a ridge known as a "splitter" which splits the oncoming high-speed jet into two equal streams. This allows for the two streams to leave the bucket in a direction which is almost opposite to that of the incoming jet. The high-speed jet is discharged to the atmosphere; therefore, all pressure drop occurs in the nozzle (stator).

Referring to Figure 5.20, the bucket's tangential speed is given by u. At the inlet the relative velocity is w_1 and the jet velocity c_1 is shown. At the exit from the bucket half of the water jet stream leaves the bucket, as shown in the velocity diagram (Figure 5.20). The relative velocity, w_2, makes an exit angle β_2 with the tangential velocity direction. The absolute velocity at the exit c_2 is determined from the vector sum of w_2 and u. Note that u does not change direction.

From Euler's turbine equation and from the velocity diagram (Figure 5.20) the energy transfer per unit mass can be written as $gH = u_1c_{u1} - u_2c_{u2}$.

For Pelton turbines, $u_1 = u_2 = u$, $c_{u1} = c_1 = u + w_1$, and $c_{u2} = u + w_2 cos\beta_2$. Because of the friction on the fluid the relative velocity will decrease by 10–20% at the outlet of the bucket. This is accounted by a factor k, hence $w_2 = kw_1$. The Euler equation can then be written including the effect of friction as:

$$gH = u\left[u + w_1 - \left(u + w_2 cos\beta_2\right)\right] = u\left(w_1 - w_2 cos\beta_2\right) = uw\left(1 - k\,cos\beta_2\right) \quad (5.33)$$

$$gH = u\left(c_1 - u\right)\left(1 - k\,cos\beta_2\right) \quad (5.34)$$

The runner efficiency η_R is obtained by dividing the energy transfer per unit mass by the kinetic energy $c_{1/2}^2$.

$$\eta_R = \frac{2gH}{c_1^2} = 2\frac{u}{c_1}\left(1 - \frac{u}{c_1}\right)\left(1 - k\,cos\beta_2\right) \quad (5.35)$$

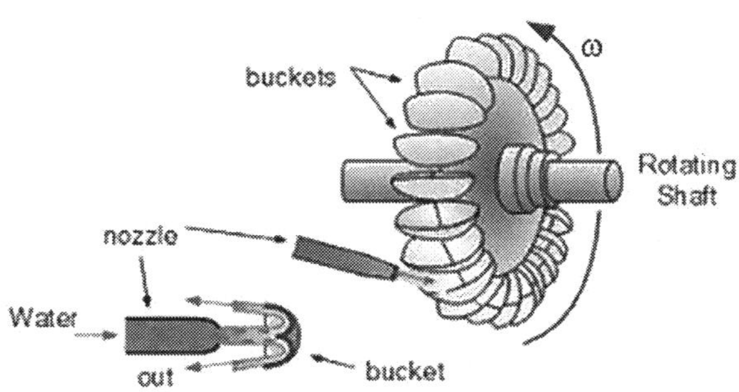

FIGURE 5.19 Pelton wheel.

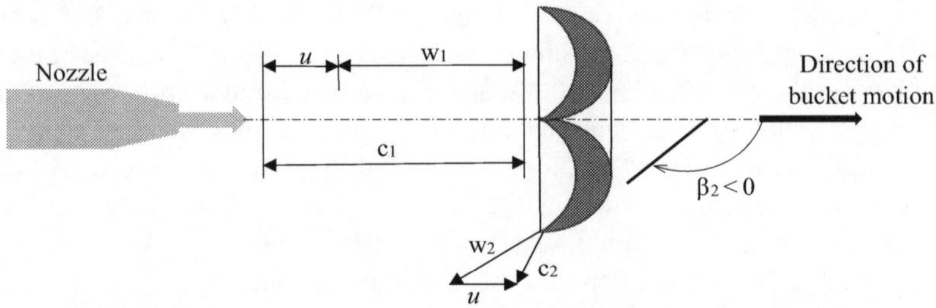

FIGURE 5.20 Velocity triangles at the inlet and outlet of a Pelton turbine bucket.

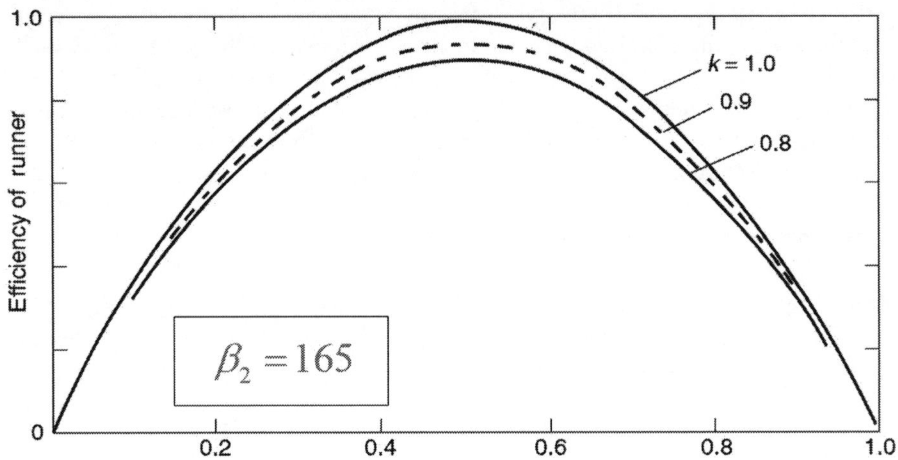

FIGURE 5.21 Blade speed/jet speed ratio (Dixon and Hall, 2013).

By defining the blade speed to jet speed ratio as $\nu = \dfrac{u}{c_1}$, the above equation can be rewritten as follows:

$$\eta_R = 2\nu(1-\nu)(1-k\,cos\beta_2) \tag{5.36}$$

Differentiating with respect to the blade speed ratio, ν, and equating the equation to zero, we obtain the maximum efficiency.

$\dfrac{d\eta_R}{d\nu} = 2\dfrac{d}{\nu}(\nu-\nu^2)(1-k\,cos\beta_2) = 2(1-2\nu)(1-k\,cos\beta_2)$. The maximum efficiency occurs when $\nu = \dfrac{u}{c_1} = 0.5$.

Figure 5.21 shows the variation of the runner efficiency with blade speed ratio for k-values of 0.8, 0.9, and 1.0 with $\beta_2 = 165°$.

The shaft power P of the Pelton turbine is determined from the above relations if hydraulic, η_h, leakage, η_v, and mechanical, η_m, efficiencies are known. Therefore, we can write:

$$P = \eta_h \eta_m \eta_v QgH \tag{5.37}$$

5.4.2 Procedure for Determining the Main Dimensions of Pelton Turbines

To determine the main dimensions of a Pelton turbine we will use the following steps. We assume that the head and flow rate available at the site are known.

5.4.3 Turbine Rotational Speed

Number of buckets, z_B, can be determined from the following relationship:

$$z_B = \frac{D_1}{2d_1} + (14 - 16) \tag{5.38}$$

where D_1 is diameter of the runner and d_1 is diameter of the nozzle. The specific speed is determined from the following empirical relationship, where H is in meters.

$$n_s = z_B\left[89.626 - 10.2\ln(H)\right] \tag{5.39}$$

$$n = \frac{n_s}{3.65}\frac{H^{0.75}}{\left(\frac{Q}{z_B}\right)^{0.5}} \tag{5.40}$$

The relationship between head, specific speed, and number of buckets is shown in Figure 5.22.

5.4.4 Determination of Runner Diameter and Nozzle Diameter

The force of the free jet is given by:

$$F = \rho Q(u_1 - c_1) = \rho a_1 c_1 (u_1 - c_1) \tag{5.41}$$

where F is the force, ρ is density of water, Q is the flow rate, u_1 is the tangential velocity of the wheel, a_1 is the cross-sectional area of the nozzle, and c_1 is the jet velocity. Taking the derivative of the above equation with respect to the jet velocity, c_1, and setting it to equal zero, the optimum tangential velocity is obtained.

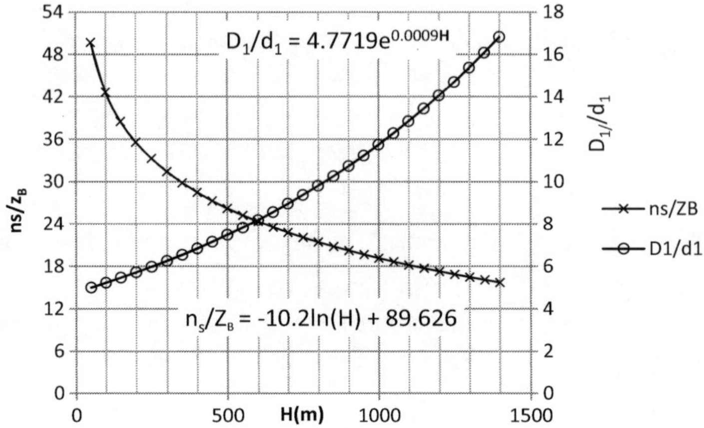

FIGURE 5.22 Relationship between slope, runnability, number of nozzles, and the ratio D_1 / d_1.

$\frac{dF}{dc_1} = \rho a_1 (c_1 - 2u_1) \rightarrow u_1 = \frac{c_1}{2} \rightarrow c_1 = 2u_1$ since $u_1 = k_u\sqrt{2gH}$, where k_u is a coefficient with the value of $k_u = 0.46–0.49$. Therefore, $c_1 = 2k_u\sqrt{2gH} = k_c\sqrt{2gH}$.

The tangential velocity of the wheel is given by $u_1 = \pi D_1 n$, resulting in $D_1 = \frac{u_1}{\pi n}$

The diameter of the nozzle is obtained from: $d_1 = \sqrt{\frac{4\frac{Q}{z_B}}{\pi c_1}}$ where k_c has a value of k_c =0.96–0.99. d_1 can also be calculated from the following equations.

$$\frac{D_1}{d_1} = 4{,}7719e^{0{,}0009H} \rightarrow d_1 = \frac{D_1}{4{,}7719e^{0{,}0009H}} \tag{5.42}$$

5.5 REACTION TURBINES

The primary features of a reaction turbine are:

- At the inlet to the turbine there is only a partial pressure drop; the rest of the pressure drop takes place in the turbine itself.
- The liquid completely fills all the spaces in and around the impeller, unlike the Pelton turbine, where the water stream is in contact with only the buckets.
- In reaction turbines rotating vanes are used for flow control.
- A draft tube is situated at the outlet of the turbine impeller; the draft tube is an integral part of the turbine.
- The water pressure gradually decreases as it flows through the impeller. This reaction to a change in pressure gave the name of a reaction turbine.

FIGURE 5.23 Francis turbine runner.

5.5.1 Francis Turbines

Designed by American scientist James Francis, Francis turbines (Figure 5.23) are the most widely used turbines in hydropower plants. Francis turbines are types of reaction turbines. Water from the penstock enters a spiral casing, which distributes water through adjustable guide vanes located around the turbine wheel. Then the water enters in a direction perpendicular to the axis of rotation, i.e., radial direction, and leaves in the axial radiation after striking and interacting with the turbine blades, hence the name mixed-flow turbine (Figure 5.11). Francis turbines are used in medium- to large-scale hydropower plants. They can be used for a wide range of heads, ranging between 2 and 300 meters. Although most Francis turbines are arranged vertically, some smaller machines can also be horizontal.

5.5.1.1 Procedure for Determining the Main Dimensions of Francis Turbines

Once the hydropower location is identified, the flow rate (Q) and the head available (H) are known. The turbine power is given by:

$$P = \rho g Q H \eta_t \tag{5.43}$$

where is water density, g is acceleration due to gravity, Q is the flow rate, H is the head, and η_t is the overall turbine efficiency. The specific speed is given by:

$$n_s = 1.166 n \frac{\sqrt{P}}{H\sqrt[4]{H}} \tag{5.44}$$

And the rotational speed of the turbine wheel is given by:

$$n = \frac{n_s H\sqrt[4]{H}}{1.166\sqrt{P}} \tag{5.45}$$

The diameter of the runner is calculated using the following equation:

$$D_{1a} = ku_{1a}.\sqrt{2gH}.\frac{60}{\pi.n} \tag{5.46}$$

where

$$ku_{1a} = 0{,}000004.n_s^2 - 0{,}0008.n_s + 0.72 \tag{5.47}$$

The number of blades is determined from the following equation:

$$z = \frac{10 \div 12}{k_{u1a}} \tag{5.48}$$

5.6 KAPLAN TURBINES

A Kaplan turbine is an axial flow reaction turbine, in which the flow direction does not change as it flows through the rotor. Kaplan turbines are propellers with adjustable blades inside a tube. In Kaplan turbines, water flows in and out along its rotational axis (axial flow). The blades of Kaplan turbines can be rotated to change their angle. This helps maintain maximum efficiency for different flow rates. Kaplan turbines are used primarily for low-head applications, i.e., for $H < 50$ m. Like Francis turbines, adjustable inlet guide vanes, located around the inside volute casing, are used to regulate the amount of flow.

5.6.1 Procedure for Determining the Main Dimensions of Kaplan Turbines

We assume that the head and the flow rate available at the site are known. We also assume the specific speed and type of turbine are known. Assuming the head, H, flow rate, Q, and the diameter of the runner, D_1, are known, we can determine the rest of the dimensions. We will use Table 5.5 to determine unit quantities and unit flow rate, from which we will obtain the other dimensions. Unit quantities of a turbine are quantities related to a turbine, which are obtained when the turbine operates under unit head (i.e., $H = 1$ m).

Unit speed is calculated using:

$$n' = \frac{nD}{\sqrt{H}} \tag{5.49}$$

Unit flow rate is calculated using:

$$Q' = \frac{Q}{\sqrt{H}D^2} \tag{5.50}$$

TABLE 5.5
Number of Half Poles for Given Rotation Speeds

***n* (rpm)**	**3500**	**1750**	**1167**	**875**	**700**	**583**	**500**	**438**	**389**	**350**	**318**	**292**
# of half poles	1	2	3	4	5	6	7	8	9	10	11	12
n (rpm)	269	250	233	219	206	194	184	175	167	159	152	146
# of half poles	13	14	15	16	17	18	19	20	21	22	23	24

The rotational speed of the turbine wheel is given by:

$$n = \frac{n_s}{3.65}\frac{H^{0,75}}{\sqrt{Q}} \tag{5.51}$$

The diameter of the runner is given by:

$$D_1 = 84.6\frac{k_{u1}\sqrt{H}}{n} \tag{5.52}$$

where n is in rpm, and k_{u1} is speed coefficient.

5.7 DRAFT TUBE ANALYSIS

A draft tube can have different shapes depending on the installation requirement. Most common are elbow and S-types. The elbow-type draft tube is shown in Figure 5.24. It consists of a straight section, i.e., straight cone, the knee itself, and finally the third part has the shape of a diffuser. A typical cross-section of the diffuser portion is a gradual transition from a circular cross-section to a rectangular cross-section, as shown in Figure 5.24.

An S-type type of draft tube is shown in Figure 5.24. The characteristic course of the surfaces from the inlet to the outlet of the knee and S-type draft tubes is shown in Figure 5.25. Such a characteristic course of the surfaces guarantees that the flow separation behind the knee will be as small as possible.

5.8 EXAMPLE PROBLEMS

Example 5.1

A hydropower site assessment yielded the following results: available head 100 m, power output 10,000 kW at rotational speed of 600 rpm. The available head at a test laboratory is 20 m and the model to the prototype scale is proposed to be 1/8. Determine (a) the speed at which the model should be tested; (b) the power input required for the model test; and (c) the model turbine flow rate.

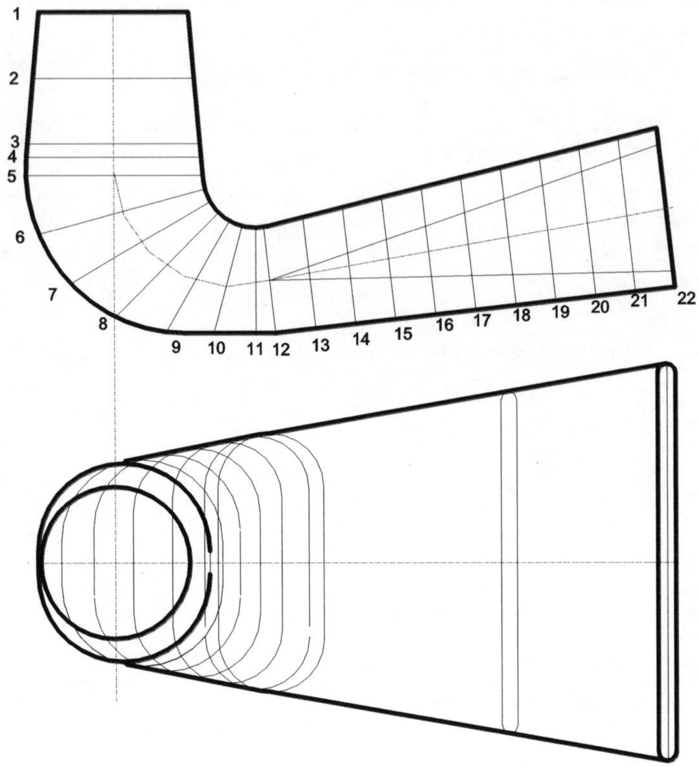

FIGURE 5.24 Elbow draft tube.

Given

$$H_p = 100m \quad P_p = 10000kW \quad n_p = 600rpm \quad \varpi_n = 62.832s^{-1}$$

$$H_m = 10m \quad \frac{D_p}{D_m} = \frac{1}{8} \quad \rho = 1000\frac{kg}{m^3}$$

(*a*)

$$\frac{H_m}{n_m^2 D_m^2} = \frac{H_p}{n_p^2 D_p^2}$$

$$n_m = \sqrt{\frac{H_m \cdot n_p^2 \cdot D_p^2}{H_p \cdot D_m^2}} = 119.215\frac{1}{s}$$

$$n_m = 1138rpm$$

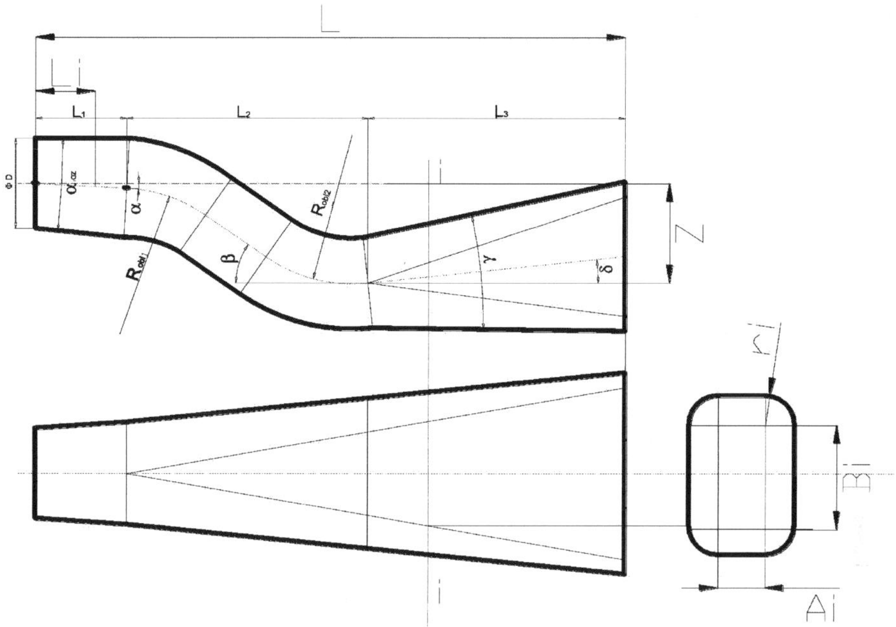

FIGURE 5.25A S-type draft tube.

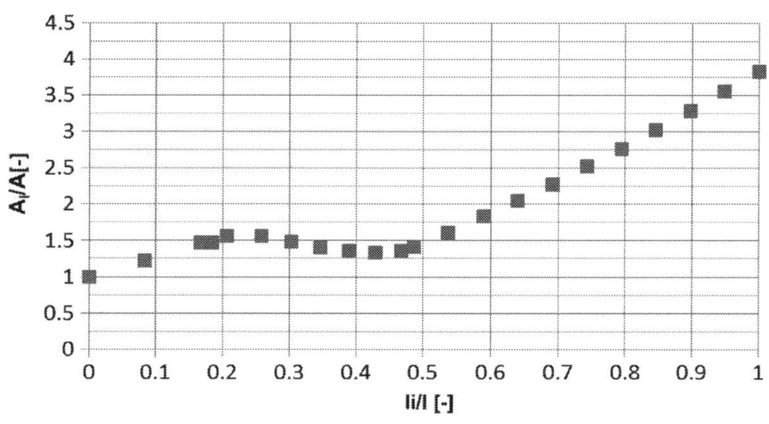

FIGURE 5.25B Characteristic course of diffusion in elbow and S-type draft tube.

(*b*)

Note that for the mod *el and prototype to be similar thier specific speeds should be the same*

$$n_{sm} = n_{sp}$$

$$\frac{n_m\sqrt{P_m}}{H_m^{\frac{5}{4}}}=\frac{n_p\sqrt{P_p}}{H_p^{\frac{5}{4}}}$$

$$P_m=\left(\frac{n_p\sqrt{P_p}H_m^{\frac{5}{4}}}{n_mH_p^{\frac{5}{4}}}\right)^2=8784W$$

(c)

$$\frac{Q_m}{n_mD_m^3}=\frac{Q_p}{n_pD_p^3}$$

$$\frac{Q_m}{Q_p}=\frac{n_m1^3}{n_p6^3}=0.009$$

The mod *el turbine flow rate is approximately* $\frac{1}{11^{th}}$ *of that of the prototype*

Example 5.2

Given the following parameters, select the appropriate turbine using Table 5.1. $H = 260$ m, $Q = 2.1$ m^3/s, for $n = 750$ rpm, and $n = 1000$ rpm. The overall efficiency is 88%.

$$P=\frac{\left(1000kg/m^3\right).\left(9.81m/s^2\right).\left(2.1m^3\right).\left(260m\right)}{1000}=4713.51kW$$

$$n_s=1.166\frac{n}{H}\sqrt{\frac{P}{\sqrt{H}}}=1.166\frac{750}{260}\sqrt{\frac{4713.51}{\sqrt{260}}}=57.497$$ **4 jet Pelton turbine**

$$n_s=1.166\frac{n}{H}\sqrt{\frac{P}{\sqrt{H}}}=1.166\frac{1000}{260}\sqrt{\frac{4713.51}{\sqrt{260}}}=76.663$$ **slow- running Francis turbine**

Example 5.3

Calculate specific speed of a turbine if the flow rate $Q = 22$ m^3/s, head $H = 125$ m, and $n = 1000$ rpm. Assume 100% efficiency, $\eta_t = 1$.

Power is given by:

$$P = \frac{(1000).(9.81).(22).(125)}{1000} = 26977.5kW$$

$$n_s = 1.166\frac{n}{H}\sqrt{\frac{P}{\sqrt{H}}} = \frac{1.166.1000}{125}.\sqrt{\frac{26977.5}{\sqrt{125}}} = 458.2$$

similarly, $n_s = 3.65n\frac{\sqrt{Q}}{H^{0.75}} = (3.65).(1000).\frac{\sqrt{22}}{125^{0.75}} = 458.2$

Example 5.4

Given $H = 920$ m, $P = 40$ MW, $n = 500$ rpm. Determine specific speed.

$$n_s = 1.166\frac{n}{H}\sqrt{\frac{P}{\sqrt{H}}} = \frac{(1.166).(500)}{920}.\sqrt{\frac{40000}{\sqrt{920}}} = 23.01$$

Example 5.5

A model turbine has a head $H = 10$ m, flow rate $Q = 2.5$ m^3/s, power, P, = 150 kW and $n = 600$ rpm. A prototype turbine is to be developed based on these data. The prototype will have a head of 4.5 m available. Determine the speed and flow rate of the prototype so that the model and prototype turbines have the same efficiency.

Solution

$$\frac{n}{n'} = \sqrt{\frac{H}{H'}} \Rightarrow n = n'^{\sqrt{\frac{H}{H'}}} = 600\sqrt{\frac{4.5}{10}} = 402.5rpm$$

$$\frac{Q}{Q'} = \sqrt{\frac{H}{H'}} \Rightarrow Q = Q'\sqrt{\frac{H}{H'}} = 2.5\sqrt{\frac{4.5}{10}} = 1.125\,{}^{m^3}\!/_{s}$$

Example 5.6

A Pelton wheel produces 3 MW under a head of 100 m and with an overall efficiency of 85%. Assuming the coefficient of velocity for the nozzle (nozzle efficiency) is 0.98, determine the diameter of the nozzle.

$$P := 3\cdot 10^6\ W \quad H := 100\ m \quad \eta_c := 0.85 \quad k_c := 0.98$$

$$P = (3\cdot 10^6)\ W \quad \rho := 1000\ \frac{kg}{m^3}$$

Velocity of the jet

$$c_1 := k_c \cdot \sqrt{2 \cdot g \cdot H} = 43.401 \frac{m}{s}$$

$$\eta_c = \frac{P}{\rho \cdot Q \cdot g \cdot H}$$

$$Q := \frac{P}{\rho \cdot Q \cdot g \cdot H} = 3.599 \frac{m^3}{s}$$

$$Q = A \cdot \upsilon = \frac{\pi \cdot d^2}{4} c_1$$

$$d := \sqrt{\frac{4 \cdot Q}{\pi \cdot c_1}} = 0.325 m$$

EXAMPLE 5.7

A Pelton turbine produces 10 MW under a head of 300 m. If the Pelton wheel is rotating at a rotational speed of 500 rpm, determine the flow rate and the jet nozzle diameter. Assuming an overall efficiency of 85%, and that the coefficient of velocity for the nozzle (nozzle efficiency) is 0.98, and $k_u = 0.46$, determine the diameter of the nozzle.

$P := 10 \cdot 10^6 \ W$ $\quad H := 300 \ m$ $\quad \eta_c := 0.85$ $\quad N := 530 \ rpm$

$P = (1 \cdot 10^7) \ W$ $\quad \rho := 1000 \frac{kg}{m^3}$ $\quad k_c := 0.98$

Velocity of the jet

$$c_1 := k_c \cdot \sqrt{2 \cdot g \cdot H} = 75.173 \frac{m}{s}$$

$$\eta_c = \frac{P}{\rho \cdot Q \cdot g \cdot H}$$

$$Q = \frac{P}{\rho \cdot Q \cdot g \cdot H} = 3.999 \frac{m^3}{s}$$

$$Q = A \cdot c_1 = \frac{\pi \cdot d^2}{4} c_1$$

$$d := \sqrt{\frac{4 \cdot Q}{\pi \cdot c_1}} = 0.26 m$$

Assume $k_u = 0.46$; k_u is between 0.46 and 0.49.

$$u_1 = k_u \cdot \sqrt{2 \cdot g \cdot H}$$

$$u_1 := 0.46 \cdot c_1 = 34.58 \frac{m}{s} \qquad \omega = \frac{(\pi \cdot N)}{30}$$

$$u_1 = \omega \cdot \frac{D}{2} \qquad \omega := 55.501 \cdot \frac{1}{s}$$

$$D := 2 \cdot \frac{u_1}{\omega} = 1.246\ m$$

Example 5.8

A turbine produces 10,000 kW under a head of 30 m, while running at 170 rpm. The diameter of the runner is 4.2 m while the hub diameter is 2 m; the discharge is 60 m^3/s. Determine: (a) the turbine efficiency; (b) specific speed; (c) type of turbine; and (d) flow area.

$$P := 15MW \qquad H := 30\ m \qquad N := 170\ rpm$$

$$D_1 := 2\,m \qquad D_2 := 4.2\,m \qquad Q := 60 \frac{m^3}{s}$$

$$N_3 = \frac{N \cdot \sqrt{P}}{H^{\frac{5}{4}}} \qquad N_s := \frac{170 \cdot \sqrt{15000}}{30^{\frac{5}{4}}} = 296.547$$

$$P_t := \rho \cdot Q \cdot g \cdot H = (1.765.10^7)\ W$$

Flow area

$$A := \pi \cdot \frac{(D_2^{\ 2} - D_1^{\ 2})}{4} = 10.713\ m^2$$

$$\eta_t := \frac{P}{P_t} = 0.85$$

Type of turbine: high-speed Francis turbine.

5.9 Exercise Problems

Problem 5.1

Assessment of a potential hydropower site shows a head of 200 m and 40,000 kW available power. The speed of the prototype turbines was determined to be 500 rpm.

A test laboratory has available head of 20 m, and a model was designed at a scale of 1/6th of the prototype. Determine the test dynamometer speed and power requirement, and the model flow rate in terms of the prototype flow rate.

Problem 5.2

Use the data for the proposed hydropower plant from the previous problem. Assume that the test laboratory has a dynamometer operating at 1000 rpm. For these values, determine the model power and the required head. The specific speed of the proposed model is 155.

Problem 5.3

A turbine operates at 400 m and $n = 500$ rpm, $Q = 5$ m^3/s and produces power $P = 17.66$ MW. If the available head is reduced to $H = 350$ m, what will be the new rotational speed, flow rate, and power?

Problem 5.4

A lawn sprinkler is shown in the figure below. The sectional area at the outlet is 1 cm^2. The flow rate is 1.5 l/s on each side. Calculate the angular speed of rotation and the torque required to hold it stationary. Neglect friction.

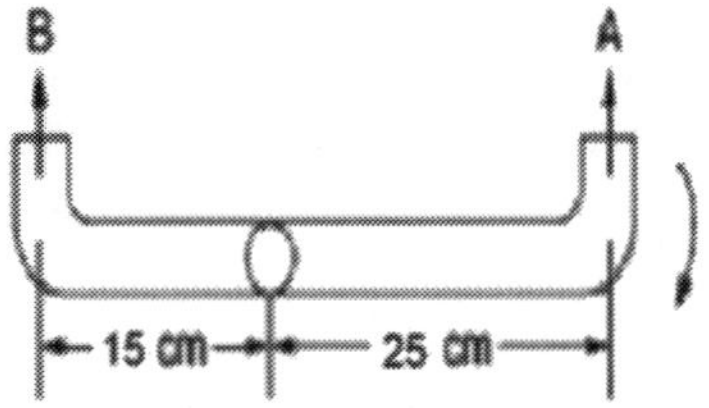

Problem 5.5

A turbine operates at $n = 500$ rpm at a head of $H = 550$ m. A jet of 20 cm is used. Determine the specific speed of the turbine and the pitch diameter of the runner. Assume $k_c = 0.97$, $k_u = 0.46$, and the overall efficiency is $\eta = 88\%$.

Problem 5.6

A Pelton turbine produces 2800 kW under a head of 400 m. If its specific speed is 21 and $k_c = 0.98$, determine the jet diameter and rotational speed.

Problem 5.7

A turbine produces 50,000 kW under a head of 30 m, while running at 180 rpm. The diameter of the runner is 4.2 m while the hub diameter is 2 m; the discharge is 150 m^3/s. Determine: (a) the turbine efficiency; (b) specific speed; (c) type of turbine; and (d) flow area.

Problem 5.8

A Pelton wheel develops 500 kW under a head of 200 m while rotating at 220 rpm. Determine the jet diameter if the hydraulic efficiency is 86%, and nozzle efficiency $k_c = 0.98$.

Problem 5.9

A Pelton turbine is designed for a head of 200 ft. The nozzle diameter, d, is 5 in; nozzle efficiency k_c is 0.98. The overall efficiency is 0.9, and the wheel rotates at $N = 150$ rpm. Determine the flow rate, shaft power, and torque.

5.10 BIBLIOGRAPHY

Bohl, W. *Fluid Flow Machines, Vol. 2, Calculation and Construction (in German)*. Munich: Vogel, 1999.

Dixon, S. L. and Hall, C. A. *Fluid Mechanics and Thermodynamics of Turbomachinery, 7th edition*. Oxford: Elsevier, 2013.

Fraenkel, P., Paish, O., Harvey, A., Brown, R., Edwards, A. and Bokalders, V. *Micro-Hydro Power: A Guide for Development Workers*. Semantic Scholar, DOI:10.3362/9781780442815, 1999.

Hlbočan, P. and Varchola, M. "Numerical simulation on a mixed-flow pump operating in a turbine mode," *Eng. Mech.*, 20, 97–105, 2013.

Kothandaraman, R. and Rudramoorthy, C. P. *Fluid Mechanics and Machinery*, second edition. New Delhi: New Age International, 2007.

Krueger, R. E. *Selecting Hydraulic Reaction Turbines*. Washington: Office of Design and Construction, Engineering and Research Center, Washington, 1976.

Melichar, K. B. and Blaha, J. *Hydraulic Machines Construction and Operation (in Czech)*. Prague: CVUT, 2002.

Nechleba, M. *Water Turbines, their Construction and Accessories (in Czech)*. Prague: SNTL, 1962.

Nechleba, M. and Hušek, J. *Hydraulic Machines (in Czech)*. Prague: SNTL / SVTL, 1966.

Paish, O. "Micro-hydropower: Status and prospects," *Proc. Inst. Mech. Eng. Part A J. Power Energy*, 216, 31–40, 2002.

Raabe, J. "Hydraulic machines and systems (in German)," in *How Hydraulic Machines Work (in German)*, 2nd edition. Düsseldorf: VDI-Verlag, 1989.

Steponoff, A. J. *Centrifugal and Axial Fiow Pumps: Theory, Design and Application*, second edition. Malabar, FL: Krieger, 1957.

Sultanian, B. K. *Logan's Turbomachinery*. Boca Raton, FL: CRC Press, 2019.

Varchola, M. "Hydro turbine for small head and high speed," in *Hydroturbo 2010: 20th International Conference "On the Use of Hydropower"*. Slovakia: Slovak Technical University (STU), 2010.

Varchola, P. H. M. and Bielik, T. "Methodology of 3D hydraulic design of a impeller of axial turbo machine," *Eng. Mech.*, 20, 107–118, 2013.

Varchola, M. and Hlbočan, P. *Hydraulic Design of an Axial Machine*. Bratislava, Slovakia: STU Bratislava, 2015.

Willi Bohl, W. E. *Fluid Flow Machines 1. Structure and Mode of Operation (Kamprath series) (in German)*. Würzburg: Vogel Communications Group.

6 Small Hydropower Plants

6.1 INTRODUCTION

According to the US Department of Energy, small hydropower plants (SHPs) are those hydropower plants that generate 10 MW or less. In some cases, SHPs are defined as hydropower plants with output of up to 15 MW for axial and Kaplan turbines and an output of 30 MW for Francis and Pelton turbines. SHPs have been identified as having great potential in rural electrification. SHPs can provide the necessary power for industrial, agricultural, and domestic use. Hence it is believed that the 1.1 billion people in rural areas of developing countries who do not have access to electricity can benefit from SHPs. According to the United Nations Industrial Development Organization (UNIDO), global SHP installed capacity has increased by an impressive 10% and keeps growing (Figure 6.1).

According to UNIDO, China continues to dominate the global SHP landscape with 54% of the world's total installed capacity. China has 28% of the world's total SHP potential, more than four times the SHP installed capacity of Italy, Japan, Norway, and the USA combined. Together, the top five countries – China, the USA, Japan, Italy, Norway, and Turkey – account for 67% of the world's total installed capacity of SHP. Although there is great potential and benefits of SHP, much of the world's SHP potential remains unexploited. The world's remaining SHP potential is almost twice that of installed capacity (Figure 6.2).

6.2 KEY FEATURES OF SMALL HYDROPOWER PLANTS

SHPs work in a similar manner as large hydropower plants. The energy of falling water is harvested to produce electricity. Depending on the site's flow rate and head, different types of turbines can be utilized. Figure 6.3 shows different types of turbine that can be selected based on the site's head and flow rate.

Similar to conventional hydropower plants, a hydro-turbine converts the energy of falling water into mechanical energy. A generator coupled to the hydro-turbine shaft converts mechanical energy into electricity. The power potential of the site mainly depends upon the head and the flow rate: $P = \rho g Q H$, where ρ is the density of water, g is acceleration due to gravity, Q is volumetric flow rate, and H is the available head. Small hydropower systems depend mainly on head and water flow to produce energy. This means that any system with flow rate and head is a potential candidate for an SHP. This includes naturally existing systems such as streams and rivers, and man-made systems such as water networks where there is elevation difference and water flow in pipes. This means that small hydropower systems can be integrated into many stages of the water supply, wastewater, and irrigation networks (Kucukali 2011). A typical run-off river SHP sketch is shown in Figure 6.4.

DOI: 10.1201/9781003007142-6

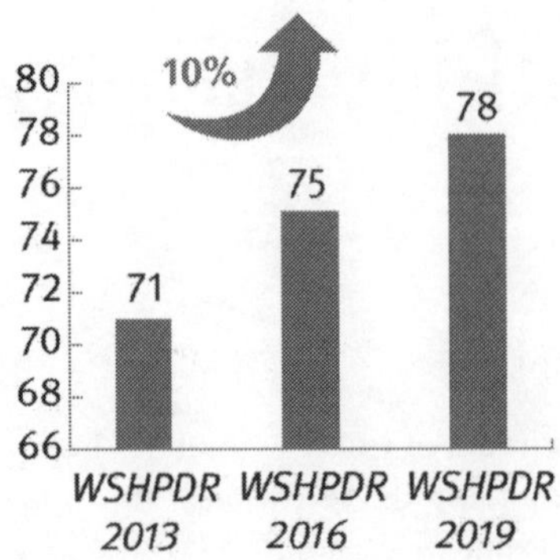

FIGURE 6.1 World small hydropower (SHP) installed capacity (GW). WSHPDR, World Small Hydropower Development Report.

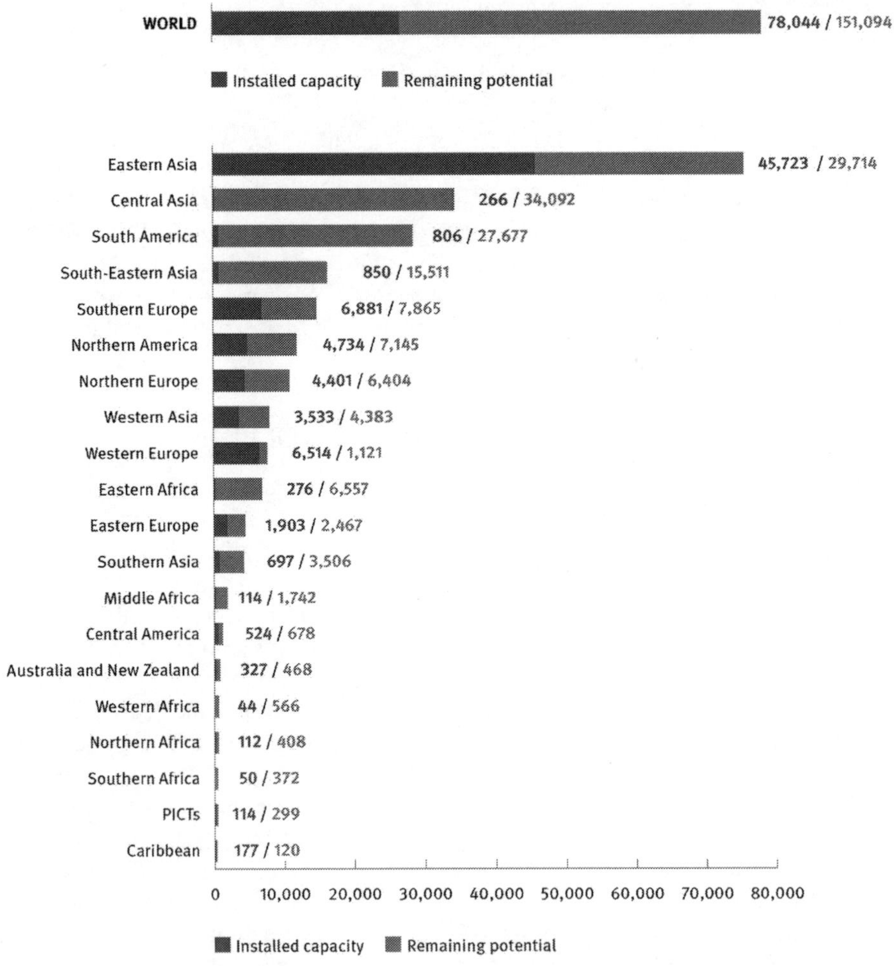

FIGURE 6.2 Remaining small hydropower (SHP) potential by region (MW). PICTs, Pacific Island Countries and Territories.

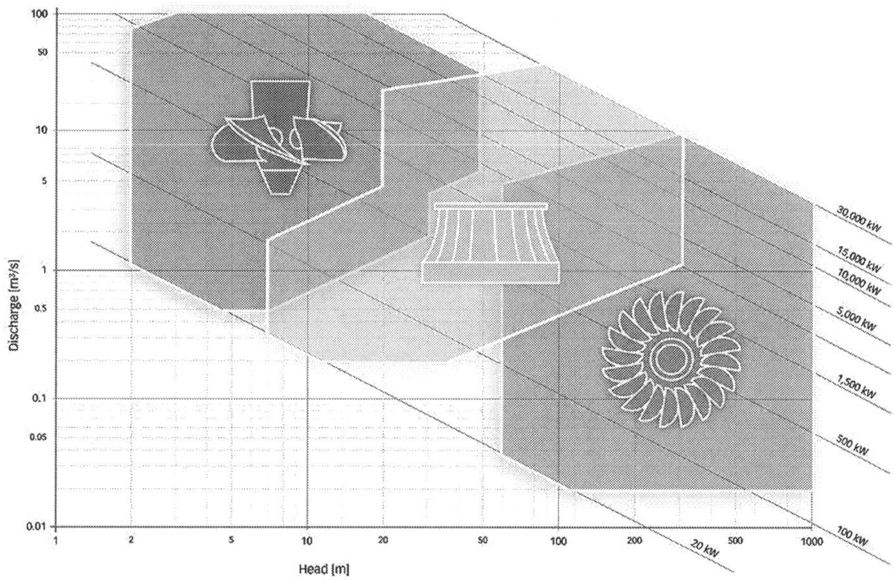

FIGURE 6.3 Types of hydro-turbines for different heads and flow rates.

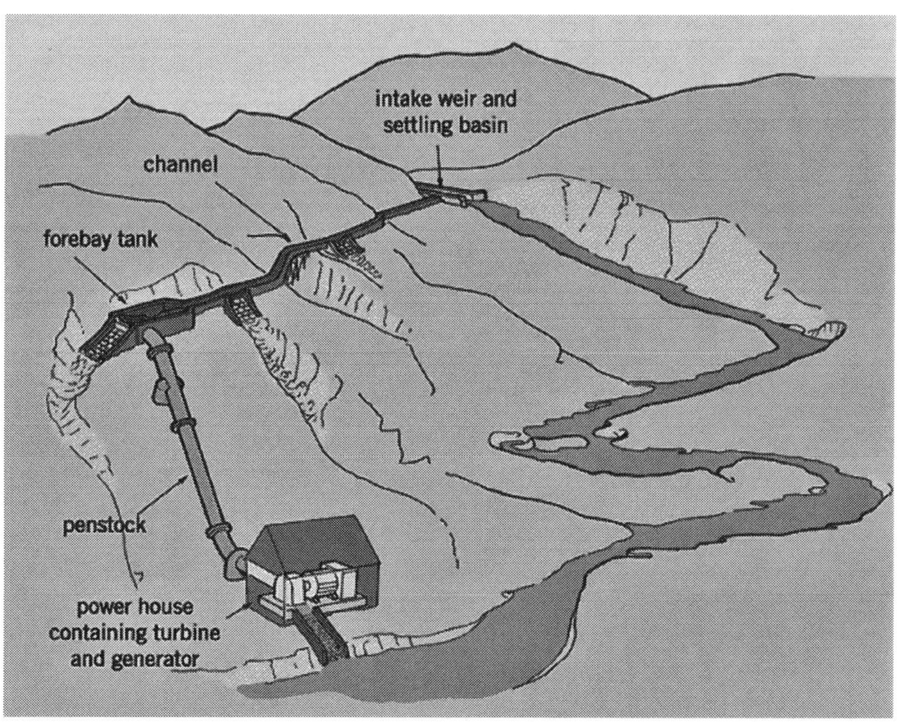

FIGURE 6.4 Sketch of typical run-of-river small hydropower plant.

6.3 FEASIBILITY STUDIES

Generally, planning a hydropower project is an iterative process. Among others, environmental impact assessment, technological options, and economic evaluation must be considered. Figure 6.5 outlines key steps in developing and planning an SHP. In most cases four phases of engineering work are usually required to develop a hydropower project at a potential site (Zhang et al. 2013):

1. reconnaissance surveys and hydraulic studies
2. pre-feasibility study
3. feasibility study
4. system planning and project engineering.

The first step in developing a hydropower system involves a site assessment. This assessment of the potential hydropower site includes site-specific information obtained from site investigations and multiple data sources are usually necessary. The site must have an adequate quantity of falling water (flow rate) and head. Although not always accessible, mountainous sites are best suited for providing higher heads. The next step is to determine the power available at any instant, using the previously described equation: $P = \rho g Q H$.

For a given site the main variables that influence the power generated are the head and the flow rate. Usually, the head for a given site remains constant and one can control the flow rate entering the turbine. However, the turbine should have the capacity to accommodate the variable flow rate. It is important to determine the head and the flow rate accurately, to assess the hydropower potential of a site. Gross head is simply the difference in elevation between the top of the penstock and the outlet from the turbine (Figure 6.6). This available head does not take into account the losses in the penstock. The net head, on the other hand, accounts for the losses in the penstock. Net head is simply the gross head minus the losses in the penstock due to friction and turbulence. Hence design of penstock should minimize these losses to maximize power output. Most small hydropower sites are categorized as low or high head. Low

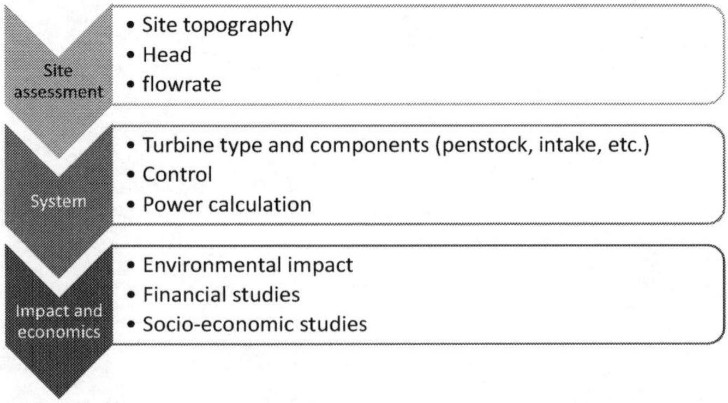

FIGURE 6.5 Planning and assessment of small hydropower plants.

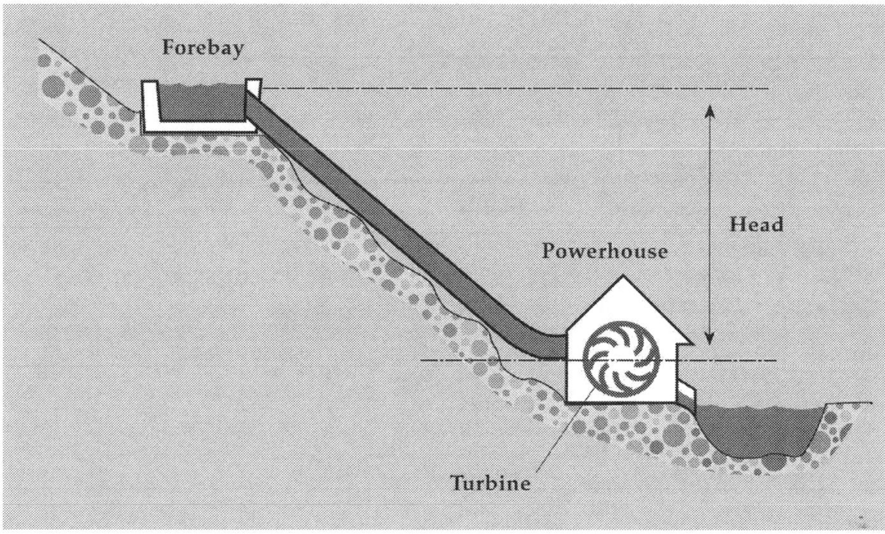

FIGURE 6.6 Determination of head available for a hydropower plant (DOE 2001).

head refers to a change in elevation of less than 10 ft (3 m) (DOE 2001). A head of less than 2 ft (0.6 m) will possibly make a small-scale hydroelectric system unfeasible (DOE 2001). Rough estimate of the head available and the flow rate can be obtained from local offices such as the US Geological Survey. To obtain a rough power output estimate, the equation described above ($P = \rho gQH$) can be used. However, we need to include the overall efficiency of the system. Assuming the overall efficiency (that accounts for hydraulic, mechanical, and volumetric losses) given by η, and the net head given by H_{net}, then the power output is given by $P = \eta\rho gQH_{net}$. Usually, an overall efficiency of 53% is assumed for SHPs (DOE 2001).

The topographical and geological assessment along with the net head, available flow rate, size of the plant, equipment (turbines, generators, etc.), and engineering works and labor determine the capital cost of a small hydropower project. It is useful to determine the cost in terms of cost per unit power which would also include projected maintenance and operating costs. Although researchers have attempted to develop empirical relationships (Aggidis et al. 2010; Gordon and Penman 1979) for estimating a cost of an SHP, it is not easy and there are no general rules. The cost is site-specific and may vary from country to country depending on regulations and incentives. Generally, the cost may be anywhere between \$1000 and –\$20,000. Maintenance costs of hydropower plants are generally relatively small compared to other technologies. SHPs can last up to 50 years or more without new major investments, meaning in the long term they are economical.

There are few software programs for evaluation of cost-effectiveness of an SHP. One of them is Mini-IDRO, free software, which can be used for technical and economical evaluation of mini hydropower plants. The software can be used to assess the discharge availability, choice of technical parameters, evaluation of energy production, benefits, and financial aspects of SHPs (http://www.seehydropower.eu/). Other available software includes HOMER and RETScreen. HOMER is a micropower software developed by the US National Renewable Energy Laboratory (NREL) to assist the design of a micro power system and to facilitate the comparison of different technologies. HOMER can model both the technical and economic factors involved in the project (homerenergy, online). RETScreen is a clean energy management software system for energy efficiency, renewable energy, and cogeneration project feasibility analysis, as well as ongoing energy performance analysis. Developed by the National Resources Canada, the software is available in 34 different languages in addition to English. The designer of an SHP is highly recommended to make use of these freely available software programs for techno-economic analyses.

6.4 DESIGN OF INTAKE AND PENSTOCKS

6.4.1 Intakes

Water intake diverts the required amount of water into a power canal or into a penstock without negative impact on the local environment, with the minimum possible head losses (ESHA 2004). Intakes can be classified according to the following criteria (ESHA 2004) (Figure 6.7):

- Power intake: the intake supplies water directly to the turbine via a penstock. These intakes are often encountered in lakes and reservoirs and transfer the water as pressurized flow.
- Conveyance intake: the intake supplies water to other waterways (power canal, flume, tunnel, etc.) that usually end in a power intake. These are most frequently encountered along rivers and waterways and generally transfer the water as free surface flow. Conveyance intakes along rivers are further classified into lateral, frontal, and drop intakes.

Intakes generally should be designed in such a way to avoid/minimize flow separation and excessive head loss through wing walls. Intakes should also be designed in such a way that they minimize head losses, which is important for SHPs. Some of the design considerations that help minimize head losses are listed below (ESHA 2004).

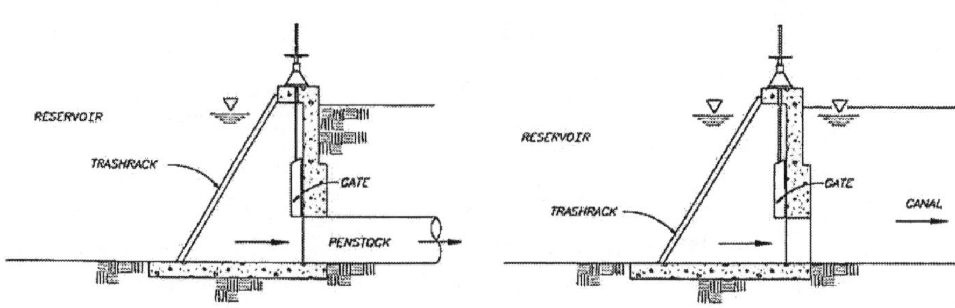

FIGURE 6.7 (a) Power intake and (b) conveyance intake (Ott 1995).

- Approach walls to the trash rack must be designed to minimize flow separation and head losses.
- Piers to support mechanical equipment, including trash racks and service gates, should be designed appropriately.
- Guide vanes should distribute flow uniformly.
- Vortex suppression devices should be used.
- Appropriate trash rack design should be utilized.

An intake should also prevent vorticity. Vorticity is undesirable because it may cause nonuniform flow conditions and introduce air into the flow. Introduction of air is undesirable because it may result in vibration, cavitation of turbine parts, unbalanced loads, increased head losses, reduced efficiency, and intake of trash and debris. Lack of sufficient submergence and asymmetrical approach seem to be the most common causes of vortex formation. Other causes of vorticity are flow separation, abrupt changes in flow direction, and approach velocities greater than 0.65 m/s (ESHA 2004).

It is also necessary to verify the minimum submergence in order to avoid vortex formation and, consequently, air entrance (Ramos 1999). Empirical relationships have been developed that determine minimum submergence required to avoid vorticity.

A criterion for intake design (Figure 6.8) in order to avoid the vortex formation is given by the following Equation [6.13]:

$$\frac{S}{D} = C\frac{V}{\sqrt{gd}} \tag{6.1}$$

where S is submergence, d is intake opening, V is the mean flow velocity at the inlet, g is acceleration due to gravity, and C is constant. $C = 1.7$ for symmetric or $C = 2.3$ for asymmetric.

According to ESHA (2004) the minimum degree of submergence is defined as shown in Figure 6.9 as h_t.

The following relationships have been developed to express the minimum values for h_t (ESHA 2004):

$$h_t \geq D\left(1 + 2.3\frac{V}{\sqrt{gD}}\right) \qquad KNAUSS \tag{6.2}$$

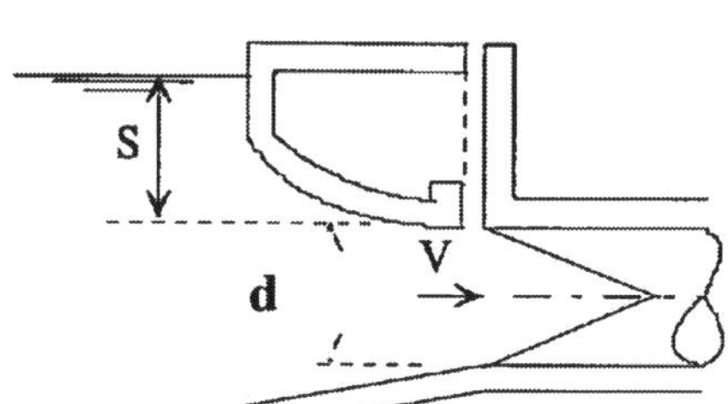

FIGURE 6.8 Schematic of the intake (Ramos 1999).

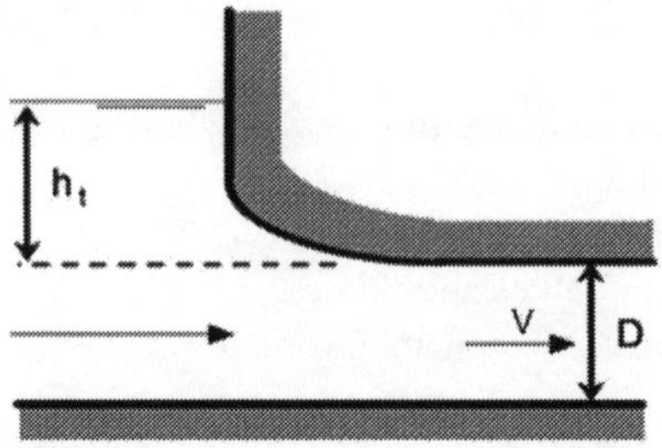

FIGURE 6.9 Minimum degree of submergence.

$$h_t \geq 4.4\left(VD^{0.5}\right)^{0.54} \quad NAGARKAR \tag{6.3}$$

$$h_t \geq 1.474V^{0.48}D^{0.76} \quad ROHAN \tag{6.4}$$

$$\begin{aligned} &h_t \geq cV\sqrt{D} \quad GORDON \\ &c = 0.7245 \text{ for asymmetric approach conditions} \\ &c = 0.5434 \text{ for symmetric approach conditions} \end{aligned} \tag{6.5}$$

Note that V is the velocity in meters per second inside the downstream conduit and D is the hydraulic diameter of the downstream conduit in meters.

6.4.2 Penstocks

A penstock transports water from the intake to the turbine. Different materials can be used for penstock, including flexible hoses, pipes made of steel/iron, and concrete. Depending on the size, penstocks can be simply laid on the ground or anchored. For larger penstocks the anchor blocks resist the thrust of the penstock. In some cases, expansion joints are needed. This depends on the temperature variation of the location and operation of the turbine (continuous or interrupted). Some of the criteria for penstock material selection are: ground conditions, site accessibility, weight, jointing system, and cost (ESHA 2004). Pipe material and size determine hydraulic losses. Smooth pipes made of plastics will minimize frictional losses. Small-diameter pipes will cause high velocity, hence high frictional losses. With increasing pipe diameter, the frictional losses decrease; however the cost will increase. Selection of penstock material requires careful hydraulic and economic analysis. Selection of the wall thickness of the penstock is usually based on the maximum internal pressure, including transient surge pressures (ESHA 2004).

As mentioned above, the diameter of a penstock is selected based on cost and head loss analysis. A rule of thumb for selection of criteria for acceptable head loss is one that results in 4% power loss. That is when the net head should result in 4% power loss when calculated using $P = \rho QgH_{net}$ from the available power using $P = \rho QgH_{gross}$.

6.4.2.1 Head Loss Calculation

The head loss in penstock comprises friction loss (h_f) due to the roughness of the inner walls of the penstock and viscous effects of the water. These losses are known as major losses. Minor losses, h_g, are due to change in the flow direction and flow area. Changes in flow direction and area are caused by, for example, control equipment such as gates or valves provided with penstock outside the reservoir/forebay to control the penstock discharge, bends provided at each location of change in penstock alignment, inlet valve provided at the end of each penstock branch to control the flow entering the turbine and transition piece, such as expansion or reducer provided before or after the inlet valve to connect it with penstock branch or with the turbine. Minor head loss also occurs at the trash rack in front of the penstock intake, which prevents floating and other materials from entering the penstock (Kumar, A. and Singhal 2015). For calculation of frictional losses, the Manning's equation can be used.

$$h_f = \frac{LV^2n^2}{R^{\frac{4}{3}}} \tag{6.6}$$

For circular penstock

$$h_f = 10.29\frac{LQ^2n^2}{D^{\frac{16}{3}}} \tag{6.7}$$

where R is hydraulic radius (ratio of the penstock cross-sectional area to its wetted perimeter), Q is flow rate, n is Manning's friction factor (Table 6.1), D is penstock diameter, and L is penstock length. Another equation that can also be used is the Darcy–Weisbach relationship.

$$h_f = f\frac{LV^2}{D2g} = f\frac{0.0826Q^2L}{D^5} \tag{6.8}$$

where f is friction factor, which can be obtained from the Moody chart for different types of relative roughness and flow regimes, V is flow velocity, and g is acceleration due to gravity.

Minor losses caused by a component integrated in the penstock (for example, a valve) can be easily calculated with the following equation:

$$h_g = K\frac{V^2}{2g} \tag{6.9}$$

where K is the loss coefficient corresponding to the component. As per the rule of thumb, if we limit the losses to be only 4%, then the diameter of the penstock can be calculated as follows:

TABLE 6.1
Manning Friction Coefficient for Different Materials (www.engineeringtoolbox.com/)

Surface material	Manning's roughness coefficient, *n*
Asbestos cement	0.011
Asphalt	0.016
Brass	0.011
Brick and cement mortar sewers	0.015
Canvas	0.012
Cast or ductile iron, new	0.012
Clay tile	0.014
Concrete – steel forms	0.011
Concrete (cement) – finished	0.012
Concrete – wooden forms	0.015
Concrete – centrifugally spun	0.013
Copper	0.011
Corrugated metal	0.022
Earth, smooth	0.018
Earth channel – clean	0.022
Earth channel – gravelly	0.025
Earth channel – weedy	0.030
Earth channel – stony, cobbles	0.035
Floodplains – pasture, farmland	0.035
Floodplains – light brush	0.050
Floodplains – heavy brush	0.075
Floodplains – trees	0.15
Galvanized iron	0.016
Glass	0.010
Gravel, firm	0.023
Lead	0.011
Masonry	0.025
Metal – corrugated	0.022
Natural streams – clean and straight	0.030
Natural streams – major rivers	0.035
Natural streams – sluggish with deep pools	0.040
Natural channels, very poor condition	0.060
Plastic	0.009
Polyethylene PE – corrugated with smooth inner walls	0.009–0.015
Polyethylene PE – corrugated with corrugated inner walls	0.018–0.025
Polyvinyl chloride PVC – with smooth inner walls	0.009–0.011
Rubble masonry	0.017–0.022
Steel – coal-tar enamel	0.010
Steel – smooth	0.012
Steel – new unlined	0.011
Steel – riveted	0.019
Vitrified clay sewer pipe	0.013–0.015

TABLE 6.1 (Continued)
Manning Friction Coefficient for Different Materials (www.engineering toolbox.com/)

Surface material	Manning's roughness coefficient, *n*
Wood – planed	0.012
Wood – unplaned	0.013
Wood stave pipe, small-diameter	0.011–0.012
Wood stave pipe, large-diameter	0.012–0.013

$$D = 2.69\left(\frac{n^2 Q^2 L}{H_{gross}}\right)^{0.1875} \tag{6.10}$$

Another important design consideration of a penstock is its wall thickness. The following equation can be used to determine the wall thickness of a penstock (Hydropower, 2004).

$$e = \frac{P_1 D}{2\sigma_f k_f} + e_s \tag{6.11}$$

where

e is wall thickness in mm

P_1 is hydrostatic pressure in kN/mm^2

D is internal penstock diameter in mm

σ_f is allowable tensile strength in kN/mm^2

e_s is extra thickness to allow for corrosion

k_f is weld efficiency

$K_f = 1$ for seamless penstock, for X-ray-inspected welds and stress relieved

$k_f = 0.9$ for X-ray-inspected welds.

It is recommended that the penstock be rigid enough to be handled without deformation in the field. It is recommended that a minimum thickness in millimeters is equivalent to 2.5 times the diameter in meters plus 1.2 mm (Hydropower, 2004).

6.5 TURBINE SELECTION (NUMBER AND TYPE)

Figure 6.3 shows that selection of turbine depends on the available flow rate and head. Another important criterion is specific speed. As discussed in Chapter 5, for low specific speed applications the Pelton wheel is appropriate, whereas reaction turbines are suitable for high specific speed applications. Cost is another factor that should be accounted for. Looking at Figure 6.3, we observe that there are overlapping regions. This means that there are more options than one to select from. In that case, all the turbines that fulfill the head and flow requirements are appropriate for the job, so the selection decision should be made based on the calculations of installed power and electricity output against costs. It should be remembered that figures such as Figure 6.3 may vary from manufacturer to manufacturer and they should be considered only as a guide.

6.6 HYDRAULIC TRANSIENTS AND DYNAMIC EFFECTS

A hydraulic transient, also known as water hammer or hydraulic shock, is a sharp pressure surge or wave produced when water flow is forced to stop suddenly or change direction abruptly (Yuce and Omer 2019). Hydraulic transient is a flow condition where the flow velocity and pressure change due to, for example, sudden flow control component changing status (for example, a valve closing). A sudden change in flow rate in a penstock will cause a great amount of water moving inside the penstock. If the system is not equipped with an adequate transient protection device such as surge tanks to overcome the transient phenomenon, it can cause the penstock to burst. A major concern should be to prevent any serious damage to the penstocks or other pressure conduits of the hydropower system from failure due to water hammer. Typically, the allowable maximum relative transient head variations $\Delta H/H$ will depend on the design head (Carravetta et al. 2018). Table 6.2 provides allowable variations for different available head ranges.

6.6.1 Preliminary Analysis

Flow disturbances are propagated in the penstock as waves with finite velocities called elastic wave velocities. The values of the elastic (pressure) wave velocities depend on both elasticity of the fluid and the penstock. The pressure wave velocity is given by [6.11]:

$$c = \sqrt{\frac{k}{\rho\left(1 + \frac{kD}{Et}\right)}} \tag{6.12}$$

where

k is bulk modulus of water (2.1×10^9 N/m^2) (Table 6.3)
E is modulus of elasticity of the penstock material (N/m^2)
t is wall thickness (mm)
ρ is density of fluid (kg/m^3).

TABLE 6.2
Maximum Allowable $\Delta H/H$ as a Function of Design Head

H (m)	$\Delta H/H$
>100	0.15–0.20
100–20	0.25–0.35
<20	0.50

Source: Carravetta et al. 2018.

TABLE 6.3
Bulk Modulus of Selected Materials (www.engineeringtoolbox.com/)

Material	Bulk Modulus, *K* (10^6 psi)	Bulk Modulus, *K* (GPa)
Aluminum, various alloys	9.9–10.2	68–70
Brass, 70-30	15.7	108
Brass, cast	16.8	116
Copper	17.9	123
Iron, cast	8.4–15.5	58–107
Iron, malleable	17.2	119
Magnesium alloy	4.8	33.1
Monel metal	22.5	155
Phosphor bronze	16.3	112
Stainless steels 18-8	23.6	163
Steel, cast	20.2	139
Steel, cold rolled	23.1	159
Steel, various	22.6–24.0	156–165

The time taken for the pressure wave to reach the valve on its return after sudden closure is known as the critical time or reflection time, which is given by $T_c = 2L/c$. For a sudden change in the flow velocity, Joukowsky found that the maximum pressure change given by [6.11] was:

$$P = c\frac{\Delta v}{g} \tag{6.13}$$

where Δv is the velocity change. Δv can be assumed to be equal to the initial flow velocity. However, if t (where t is valve closure time) is greater than T_c, then the pressure wave reaches the valve before the valve is completely closed, and the overpressure will not develop fully, because the reflected negative wave arriving at the valve will compensate for the pressure rise.

6.7 ELECTRICAL EQUIPMENT CONSIDERATIONS

The rotational speed of a turbine is directly linked to its specific speed, net head, and flow rate. In the small hydropower projects standard generators should be installed when possible, so during turbine selection it must be considered that the generator, either coupled directly or through a speed increaser to the turbine, should reach the synchronous speed, as given in Table 6.4 [E. S. A.-E. Hydropower, "Guide on How to Develop a Small Hydropower Plant," *Eur. Small Hydropower Assoc.*, 2004]. The synchronous speed refers to the rotational speed of the magnetic field in the motor's stator winding. It refers to the rate at which the alternating machine generates electromotive force. The synchronous speed is determined by the following relationship:

TABLE 6.4
Generator Synchronous Speed

	N_s for	
Number of Poles, *P*	***f* = 50Hz**	***f* = 60 Hz**
2	3000	3600
4	1500	1800
6	1000	1200
8	750	900
10	600	720
12	500	600
14	428	540

$$N_s = \frac{120f}{P} \tag{6.14}$$

$$f = \frac{PN}{120} \tag{6.15}$$

where

N_s is the synchronous speed
P is the total number of field poles
f is the frequency of the generated voltage in hertz
N is the speed of the field in revolutions per minute (rpm).

The synchronous speed calculated for different number of field poles is given in Table 6.4.

Direct connection is the best method when the turbine and generator run at the same speed and their shafts can be aligned. There are essentially no power losses, and maintenance is negligible. Turbine manufacturers will propose either a rigid or flexible coupling; however a flexible coupling that can withstand some misalignment is normally recommended [E. S. A.-E. Hydropower, "Guide on How to Develop a Small Hydropower Plant," *Eur. Small Hydropower Assoc.*, 2004]. However, in many cases, especially for low-head sites, turbine speeds will be low. Hence speed increasers are necessary to meet the high rpm of the generators.

The turbine's mechanical energy is converted to electrical energy via the electric generator. The rotor and stator are the two most important parts of the generator. The rotor is a revolving assembly to which the turbine shaft's mechanical torque is applied. A voltage is induced in the stationary component, the stator, when the rotor is magnetized. Except when coupled with a speed increaser, the generator's speed is dictated by the turbine option. Factors like turbine type and orientation influence the generator's placement and orientation. The generator for a bulb-type turbine, for example, is housed within the bulb itself (Figure 6.10b). Kaplan turbines require a vertical shaft generator with a thrust bearing (Figure 6.10a).

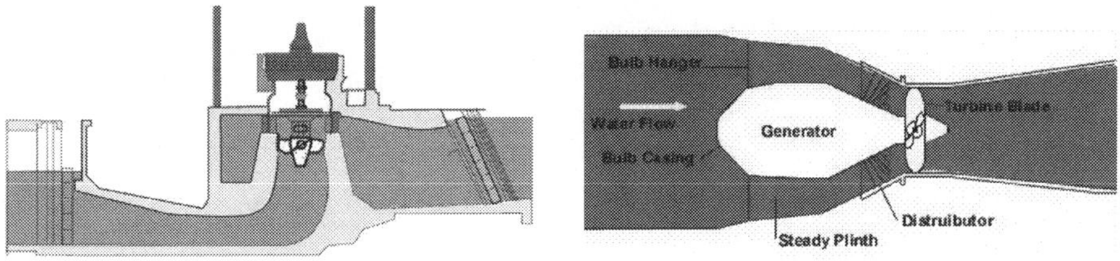

(a) (b)

FIGURE 6.10 Examples of arrangements of generators: (a) Kaplan turbine with a generator; (b) bulb turbine with a generator (www.cchpe.net/).

The generator should have a kilowatt rating that matches the turbine's kilowatt rating. Francis, fixed-blade propeller, adjustable-blade propeller (Kaplan), Pelton, and crossflow are the most prevalent turbine types. Each turbine type has unique operating characteristics, necessitating a unique set of generator design requirements to ensure that the generator is properly matched to the turbine. Regardless of turbine type, the generator must have enough continuous capacity to handle the turbine's maximum kilowatt at 90% gate opening without exceeding the generator's rated nameplate temperature rise. Any potential future adjustments to the project, such as raising the forebay (draw-down) level and boosting turbine output capabilities should be considered when evaluating generator capacity (Kumar and Singal 2013).

6.8 EXAMPLE PROBLEMS

Example 6.1

Determine the diameter of a 200-m-long penstock made of cast iron, wheref the available gross head is 100 m and the flow rte is 2.5 m³/s. It is desired that the losses due to friction do not exceed 4%.

$D = 2.69\left(\frac{n^2Q^2L}{H_{gross}}\right)^{0.1875} = 2.69\left(\frac{0.012^2 \times 2.5^2 \times 200}{100}\right)^{0.1875} = 0.822\ m$. The diameter of the pipe is taken as 1 m.

Example 6.2

Determine the frictional losses for the penstock in Example 6.2.

$$h_f = 10.29\frac{LQ^2n^2}{D^{\frac{16}{3}}} = 10.29\frac{200 \times 2.5^2 \times 0.012^2}{1^{\frac{16}{3}}} = 1.852\ m$$

Example 6.3

Determine the pressure wave velocity, for instant closure, in a penstock made of cast stainless steel of 200-mm diameter with wall thickness of 3 mm. Compare the result to that of penstock made of PVC having the same diameter and 12-mm wall thickness. Determine the surge pressure in both penstocks if the initial flow velocity is 1.2 m/s.

Solution

(a) Cast stainless steel

$$E = 139\times10^9 \frac{N}{m^2} \qquad D = 200\ mm \quad t = 3\ mm \quad \rho = 1000\frac{kg}{m^3} \qquad k = 2.1\times10^9 \frac{N}{m^2}$$

$$c = \sqrt{\frac{k}{\rho\left(1+\frac{kD}{Et}\right)}} = \sqrt{\frac{2.19\times10^9}{1000\left(1+\frac{2.19\times10^9\times0.2}{139\times10^9\times0.003}\right)}} = 1023\frac{m}{s}$$

$$P = c\frac{\Delta V}{g} = 1023\frac{1.2}{9.81} = 125.163\ m$$

(b) PVC

$$E = 2.75\times10^9 \frac{N}{m^2} \qquad D = 200\ mm \quad t = 12\ mm \quad \rho = 1000\frac{kg}{m^3} \qquad k = 2.1\times10^9 \frac{N}{m^2}$$

$$c = \sqrt{\frac{k}{\rho\left(1+\frac{kD}{Et}\right)}} = \sqrt{\frac{2.19\times10^9}{1000\left(1+\frac{2.19\times10^9\times0.2}{2.75\times10^9\times0.012}\right)}} = 391.127\frac{m}{s} = 1227kPa$$

$$P = c\frac{\Delta V}{g} = 391.127\frac{1.2}{9.81} = 47.861\ m = 469.352\ kPa$$

The example illustrates that the pressure surge in the cast stainless steel is 2.615 times higher than in the PVC. This is because cast stainless is more rigid than the PVC penstock.

6.9 BIBLIOGRAPHY

Aggidis, G. A., Luchinskaya, E., Rothschild, R. and Howard, D. C. "The costs of small-scale hydro power production: Impact on the development of existing potential," *Renew. Energy*, 35(12), 2632–2638, 2010.

Carravetta, A., Houreh, S. D. and Ramos, H. M. *Pumps as Turbines: Fundamentals and Applications*. Switzerland: Springer, 2018.

Cheng, L., Liu, C., Luo, C., Zhou, J. R. and Jin, Y. "Research on the unstable operating region of axial-flow and mixed flow pump," in *IOP Conference Series: Earth and Environmental Science*. Beijing: IOP Publishing, 15(3):2050, 2012.

DOE/GO-102001-1173, *Small Hydropower Systems*. Washington, DC: DOE, 2001.

ESHA. Hydropower, "Guide on how to develop a small hydropower plant," *Eur. Small Hydropower Assoc.*, 16, 2004.

Gordon, J. L. and Penman, A. C. "Quick estimating techniques for small hydro potential, *Int. Water Power Dam Constr.*, 31 (9), 46–51, 1979.

Kucukali, S. "Risk assessment of river-type hydropower plants by using fuzzy logic approach," in *Proceedings of the World Renewable Energy Congress – Sweden, 8–13 May, 2011*. Linköping, Sweden: Linköping University, 2011.

Kumar, A. and Singhal, M. K. "Optimum design of penstock for hydro projects," *Int. J. Energy Power Eng.*, 4 (4), 216–226, 2015.

Kumar, A. and Singal, S. K. Standards/Manuals/Guidelines for Small Hydro Development. Uttarakhand: Alternate Hydro Energy Centre, Indian Institute of Technology Roorkee, 2013.

Liu, D., Liu, H., Wang, X. and Kremere, E. (eds.), *World Small Hydropower Development Report 2019: Case Studies*. 2019. United Nations Industrial Development Organization; International Center on Small Hydro Power. [Online]. Available: www. smallhydroworld. org.

Ott, R. *Guidelines for Design of Intakes for Hydroelectric Plants*. New York, NY: American Society of Civil Engineers, 1995.

Ramos, H. *Guidelines for Design of Small Hydropower Plants*. Belfast: WREAN (Western Regional Energy Agency & Network) and DED (Department of Economic Development), 1999.

"Technical/economical evaluation of SHP plants: SMARTMini-Idro tool." [Online]. Available: www.seehydropower.eu/.

"The HOMER Microgrid Software." [Online]. Available: www.homerenergy.com/products/software.html.

Yang, K. F., Feng, J. J., Zhu, G. J., Lu, J. L. and Luo, X. Q. "Study on improvement of hump characteristic of an axial flow axial pump by grooving inlet wall," in *IOP Conference Series: Earth and Environmental Science*. Banda Aceh, Indonesia: IOP Publishing, 163(1):012090, 2018.

Yuce, M. I. and Omer, A. F. "Hydraulic transients in pipelines due to various valve closure schemes," *SN Appl. Sci.*, 1, 1110, 2019.

Zhang, Q. F. K., Uria-Martinez, R. and Saulsbury, B. *Technical and Economic Feasibility Assessment of Small Hydropower Development in the Deschutes River Basin*. UT-BATTELLE, ORNL/TM-2013/221, 2013.

7 Cavitation

7.1 INTRODUCTION: MAIN FEATURES OF CAVITATING FLOW

Cavitation occurs when the local pressure of a flowing liquid falls below the vapor pressure of the liquid. During cavitation, small bubbles or cavities filled with vapor develop, which unexpectedly collapse when the flow moves ahead into high-pressure areas. As shown in Figure 7.1, cavitation results in phase change (from liquid to vapor). The phase change can be obtained by decreasing the pressure at a relatively constant temperature and causing cavitation. Water at 10°C, for example, will vaporize and produce bubbles when the pressure decreases below 1.23 kPa. This can happen, for example, on the suction side of pumps. Cavitation appears to be identical to boiling at first glance, with the exception that the driving mechanism is different.

The driving mechanism of cavitation is the local pressure, which is regulated by flow dynamics. From a purely theoretical point of view, the cavitation process can be described by the following three stages:

1. bubble generation (cavity)
2. gradual deformation of the bubble
3. extinction of the bubble.

The effects of cavitation may be desirable or undesirable. Undesirable effects reduce the performance of a machine, cause vibrations and noise, and cause erosion of the material in places of cavitation. Desirable use of cavitation include cleaning the surface, sterilization of contaminated surgical instruments, and breakdown of pollutants

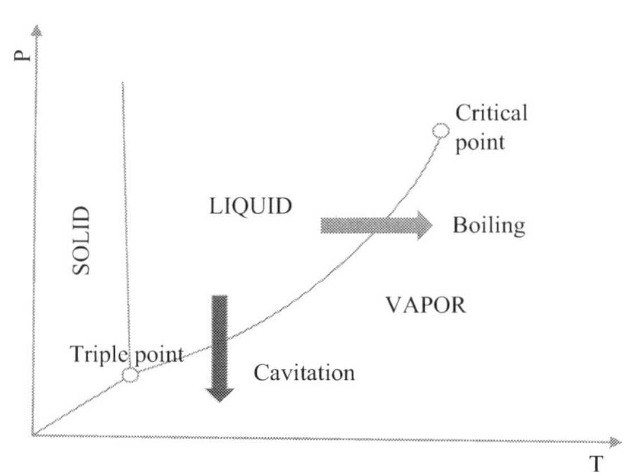

FIGURE 7.1 A typical phase diagram.

DOI: 10.1201/9781003007142-7

in water purification systems. In turbomachinery, the undesirable consequence of cavitation is cavitation erosion of functional parts. Consequently, cavitation reduces service life through rapid wear. Local pitting of rotor impeller blades due to cavitation is shown in Figure 7.2.

Cavitation effects and factors affecting their intensity are given in the Figure 7.3. Factors influencing the origin and intensity of cavitation are shown in the upper half of the boxes. In the lower boxes, the consequences of the cavitation action on the hydraulic machine are given.

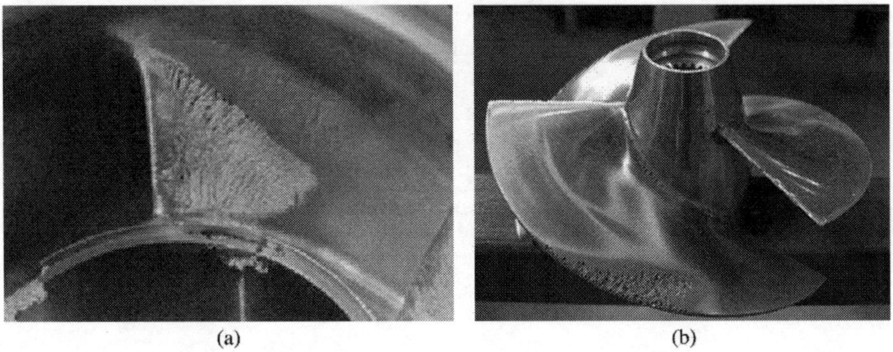

FIGURE 7.2 (a) Damage patterns on the first-stage impeller after 1 year of operation; (b) damage patterns on an axial flow pump.

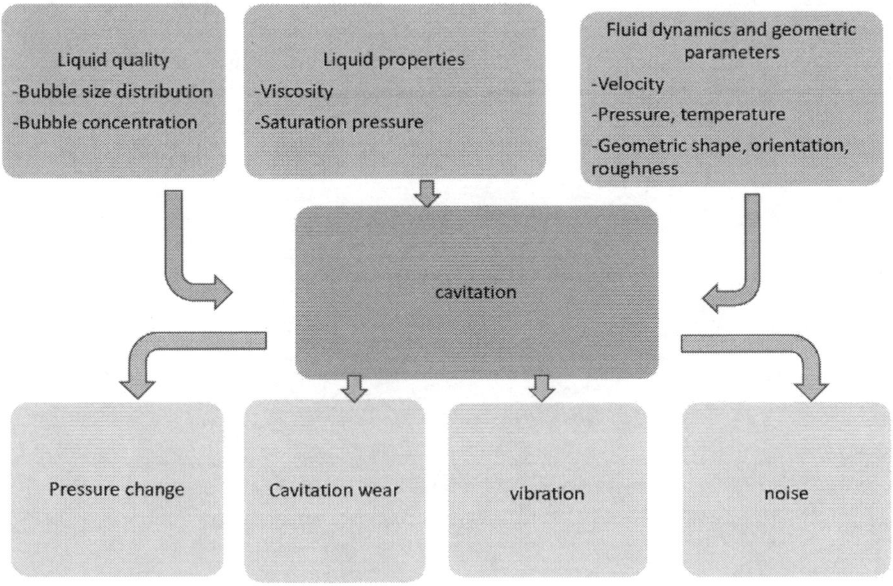

FIGURE 7.3 Block diagram of influences and effects of cavitation.

7.2 CAVITATION IN HYDRAULIC MACHINES

During operation of a hydraulic machine, the fluid flows through the parts of the machine and its flow is governed by the fundamental laws of fluid mechanics (Navier–Stokes equation and continuity equation). This means that at each point in the machine the fluid has a certain velocity and pressure. The velocity and pressure depend on the geometrical shapes of the machine part, conditions in the suction part (pressure at the machine inlet), flow rate, and pump speed. If the pressure drops below the vapor pressure value at a given point, the flow field is not continuous and cavitation bubbles are formed. This means theoretically that cavitation can occur in any part of the machine, but most often it occurs in the suction and inlet parts.

7.2.1 Cavitation in Pumps

Pumps are intended to function with a continuous flow of water, but a flooded intake may not be enough to sustain the pressure necessary to avoid cavitation. The suction side, or intake, of a pump is the place where the pressure is lowest. Flow disruptions can be caused by a variety of factors, ranging from system design to component failure. The most common reasons for flow interruption resulting in cavitation are:

- long inlet pipe
- fluid viscosity is higher than predicted
- clogged inlet
- clogged strainers and filters
- pump with a poor specification.

Long-term cavitation has obvious consequences on the pump impeller and other components. Typical signs of cavitation (Figure 7.4) include:

- noise and vibration
- failure of a seal or a bearing
- impeller deterioration (Figure 7.2)
- power usage is higher than normal.

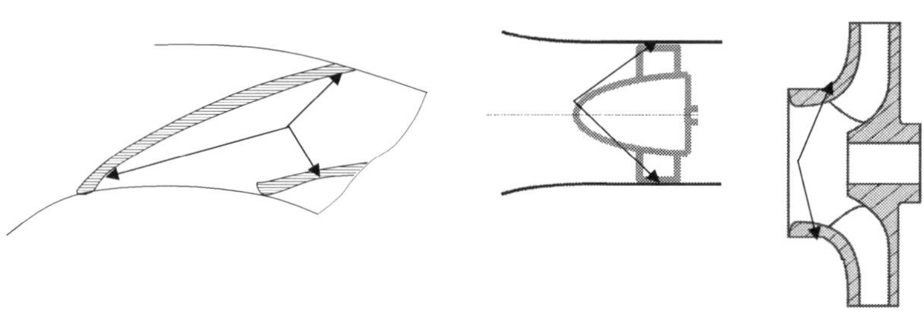

(a) (b) (c)

FIGURE 7.4 Places of cavitation are shown with arrows; at the (a) impeller blades; (b) blade tip; (c) inlet to the pump.

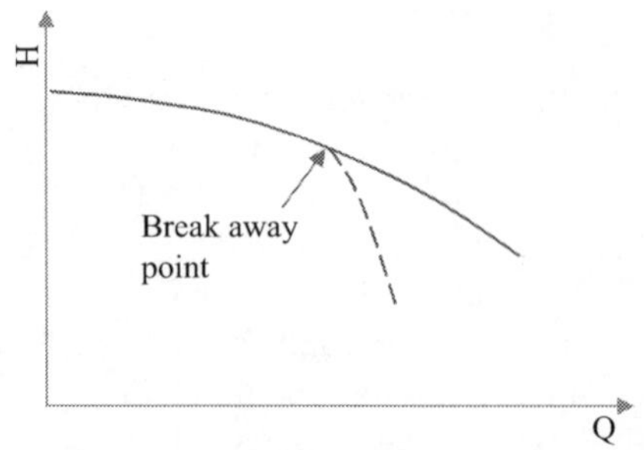

FIGURE 7.5 Effect of cavitation on the head curve of a pump.

If the minimum pressure value falls below the saturated vapor pressure, the liquid evaporates, disrupting the continuity of the liquid column and the pumping process itself. Cavitation affects the performance characteristics of the machine. Figure 7.5 shows the effect of cavitation on the performance of a centrifugal pump.

To prevent cavitation, the local pressure within the hydraulic machine should always be kept above the vapor pressure. Because pressure is easiest to measure (or estimate) at the pump's input, cavitation requirements are usually stated there. Net positive suction head (NPSH) is a flow parameter that is defined as the difference between the pump's input stagnation pressure head and the vapor pressure head.

$$NPSH = H_t - H_{vap} \tag{7.1}$$

where

H_t is the sum of the absolute static pressure head, the dynamic pressure head
H_{vap} is the absolute vapor pressure of liquid in terms of head

$$NPSH = \left(\frac{P_s}{\rho g} + \frac{V_s^2}{2g}\right)_{pump\ inlet} - \frac{P_v}{\rho g} \tag{7.2}$$

where

P_s is suction pressure
P_v is vapor pressure of the liquid
V_s is suction velocity (V_s is sometimes neglected in the literature)
ρ is density.

The aim of NPSH is to find and eliminate operating circumstances that cause the fluid to vaporize as it reaches the pump, a phenomenon known as cavitation (or flashing). The fluid pressure near the eye of the impeller of a centrifugal pump is at its lowest point. Bubbles are created when the pressure here is less than the fluid's vapor pressure, and they travel through the impeller blades to the discharge pipe. As the vapor bubbles travel through this greater pressure zone, they may spontaneously burst, causing damage to the impeller (Figure 7.6A). Figure 7.6B shows the development and disappearance (burst) of bubbles.

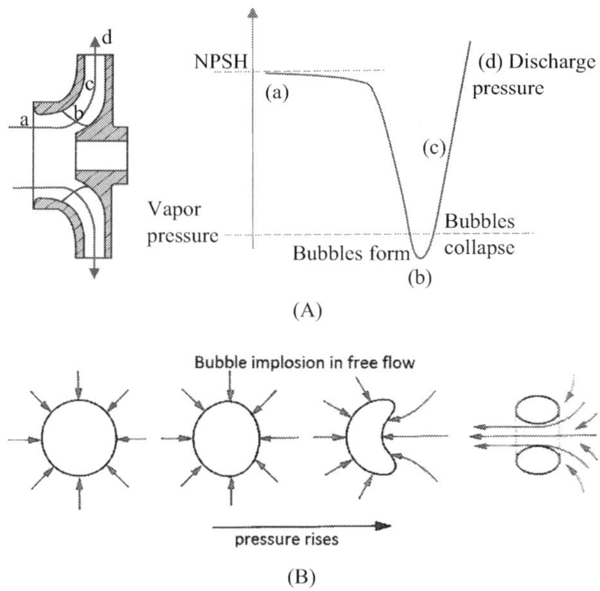

FIGURE 7.6 (A) Pressure change through a centrifugal pump experiencing cavitation: (a) fluid enters the pump; (b) pressure drops below vapor pressure at the eye of the impeller; (c) bubbles condense and collapse; (d) pressure rises as fluid leaves the impeller. (B) Development and disappearance of bubbles.

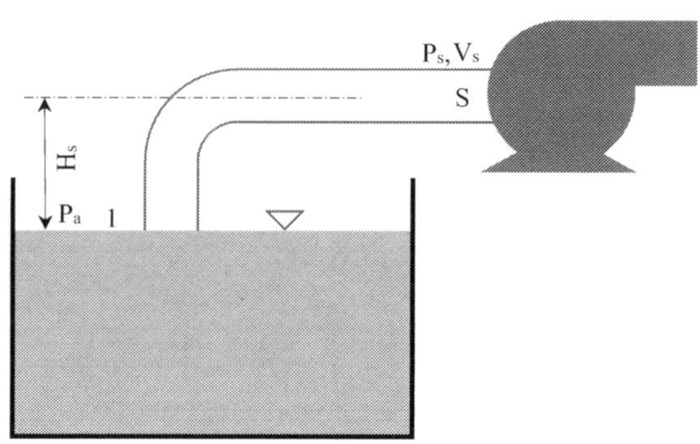

FIGURE 7.7 Suction side of pump.

We distinguish between the required and available NPSH (Figure 7.7). The required NPSH ($NPSH_R$) is a performance parameter which is the minimum NPSH necessary to avoid cavitation. $NPSH_A$ is a system function that must be calculated by the design engineer, whereas $NPSH_R$ must be supplied by the manufacturer.

Writing the Bernoulli equation for point 1 (free surface) and inlet point S, we get:

$$\frac{P_a}{\rho g} = \frac{P_s}{\rho g} + \frac{V_s^2}{2g} + H_s + h_f \tag{7.3}$$

where

P_a is atmospheric pressure

h_f is losses in the suction pipe.

Combining Equations 7.2 and 7.3, we get:

$$NPSH_A = \frac{P_a}{\rho g} - \frac{P_v}{\rho g} - H_s - h_f \tag{7.4}$$

Another useful empirical relationship used for pumps that takes into account the effect of viscosity and surface tension is given below.

$$\sigma_p = \sigma_p^* K_n K_{D2} \tag{7.5}$$

$$\sigma_p^* = 2.79 n_b^{\frac{4}{3}} \quad \text{for} \quad 0.049 \le n_b \le 0.07 \tag{7.6}$$

$$\sigma_p^* = 1.15 n_b \quad \text{for} \quad n_b \ge 0.07 \tag{7.7}$$

where

$$n_b = N \frac{\sqrt{Q}}{(g.H)^{\frac{3}{4}}} \tag{7.8}$$

N is rotational speed in s^{-1}

Q is in m^3/s

g is acceleration due to gravity in m/s^2

H is head in m

σ_p is Thoma's cavitation coefficient.

Given that cavitation is affected by viscosity and surface tension, correction factors K_n and K_D are introduced to describe the effect of Reynolds and Weber numbers:

$$K_n = 0.76 + 1.725 tg^{2.25} (60 - N) \quad for \quad 10 < N < 60 \tag{7.9}$$

$$K_{D2} = 0.78 + 9.0 tg^{2.8} \left[(0.4 - D_2) 100 \right] \quad for \quad 0.1 < D_2 < 0.4 \tag{7.10}$$

where

t is temperature

g is acceleration due to gravity

D_2 is outer diameter of the impeller.

Note that vapor pressure of a liquid is dependent on temperature. Table 7.1 gives vapor pressure of water for selected temperatures.

Note that we have designated NPSH as $NPSH_A$. The available NPSH ($NPSH_A$), which is a design parameter, must be greater than or equal to required NPSH to

TABLE 7.1
Dependence of Saturated Water Vapor on Temperature

$t(^oC)$	$p_v(kPa)$	$t(^oC)$	$p_v(kPa)$
0.0	0.608	105	120.07
38.2	6.67	129.5	266.8
51.7	13.34	143.9	400.2
66.5	26.88	154	533.6
75.9	40.80	162	667.0
83.0	66.61	193	1334.0
95.7	86.72	228	2667.3
100.0	101.33	292	6660.0
101.4	106.63	333	22092

prevent cavitation. Required NPSH is calculated by the pump manufacturer and is often included on the performance curve (Figure 7.8).

It is worth noting that, because vapor pressure is a function of temperature, the value of NPSH depends not just on flow rate but also on liquid temperature. Figure 7.8 shows that the flow rate varies with the NPSH. It also shows the sketch of NPSH which decreases with increasing flow rate. The intersection between NPSH and $NPSH_R$ gives the maximum flow rate, which is the flow rate at which the pump can operate without cavitation. It should be noted that reducing flow rate on the suction side would result in reduced suction pressure, making the pump prone to cavitation. This is one reason why flow control valves are not installed on the pump's suction side.

7.2.2 Cavitation in Turbines

Increasing the power of water turbines is achieved by increasing the speed or flow rate. The maximum specific speed of each type of turbine is limited by cavitation. After the original successful design and development of high-speed turbines such as

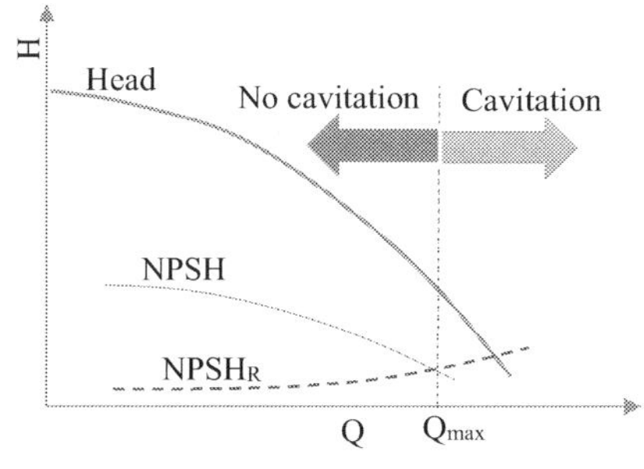

FIGURE 7.8 Illustration of the required net positive suction head ($NPSH_R$).

Kaplan, problems arose that were eventually identified as cavitation, which at that time was not sufficiently known for water turbines. Cavitation phenomena, together with their effects, have become among the main limiting factors in the performance of the turbine and the further increase in speed.

7.2.2.1 Suction Head of Water Turbines

To evaluate the cavitation performance of turbines the turbine installation with respect to the free surface should be considered. The vertical distance between the turbine and the free surface, known as suction height, H_s, is important in determining whether cavitation can occur or not (Figure 7.9).

The necessary suction head to prevent cavitation is then determined from the following relation:

$$H_s < \frac{P_a - P_v}{\rho g} - \sigma_p H \tag{7.11}$$

where

H is the turbine head
P_a is atmospheric talk
P_v is the vapor pressure of the liquid
σ_p is Thoma's cavitation coefficient.

Thoma's cavitation coefficient will be discussed later in detail. In order to prevent cavitation, the magnitude of pressure at this point must not fall below the value of the saturated vapor pressure at a given temperature. Considering Figure 7.10, the lowest pressure will be inside the impeller in the passage between the blades.

The pressure inside the blade passage will increase with increasing flow rate. Thoma's coefficient σ_p will have the highest value at the maximum flow rate through the turbine. The suction height also depends on the atmospheric pressure and thus also on the altitude (h) of the location where the turbine is to be installed. A simple equation that relates to atmospheric pressure in terms of head to altitude considering the saturated vapor of the liquid is given below:

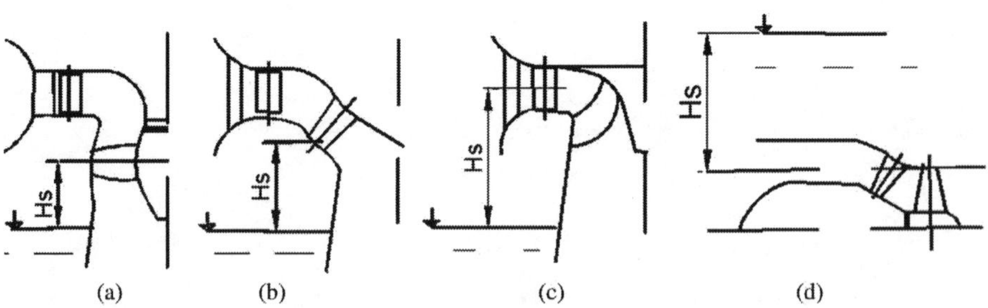

FIGURE 7.9 Measuring suction head for different types of turbines: (a) Kaplan; (b) mixed-flow; (c) Francis; and (d) bulb turbine.

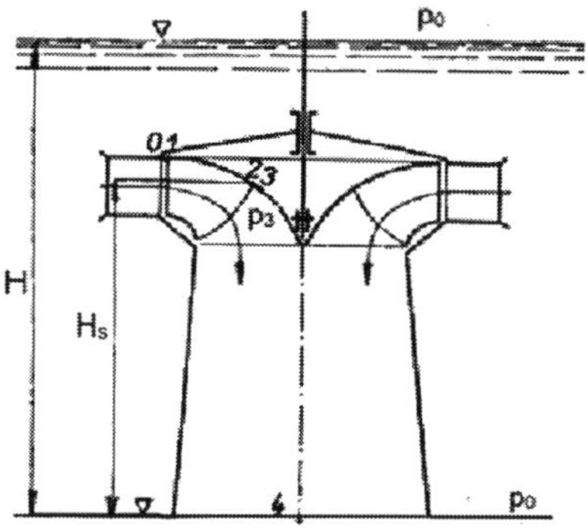

FIGURE 7.10 Water turbine draft tube.

$$H_a = 10 - \frac{h}{100} \tag{7.12}$$

Then the suction head is calculated with the following equation:

$$H_s = H_a - \sigma_p H \tag{7.13}$$

where

- H_a is atmospheric pressure (in meters)
- h is altitude above sea level (in meters).

Thoma's coefficient σ_p can be read from Table 7.2 or Figure 7.11. Thoma's cavitation coefficient depends mainly on the type of turbine, i.e., on its specific speed. Approximate values of this coefficient are given in Table 7.2 and Figure 7.11.

7.2.2.2 Exit Velocity

Considering Figure 7.12, we can write the Bernoulli equation for two points, at the exit of the turbine and the free surface.

$$\frac{P_e}{\rho g} + \frac{V_e^2}{2g} + H_s = \frac{P_a}{\rho g} + h_f \tag{7.14}$$

$$V_e = \sqrt{2\left[\frac{(P_a - P_e)}{\rho} - g(H_s - h_f)\right]} \tag{7.15}$$

TABLE 7.2
Values of σ_p on the Specific Speed

n_s (rpm)	51	102	204	299	409	511	519	701	818
σ_p	0.02	0.05	0.11	0.2	0.35	0.59	0.91	1.45	2.15

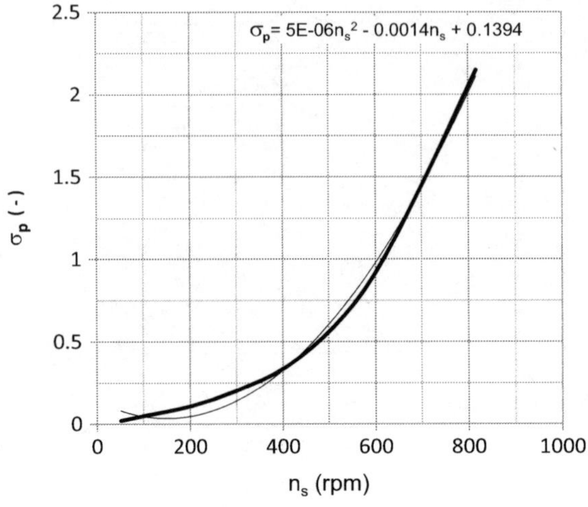

FIGURE 7.11 Values of cavitation: Thoma's coefficient for different specific speeds.

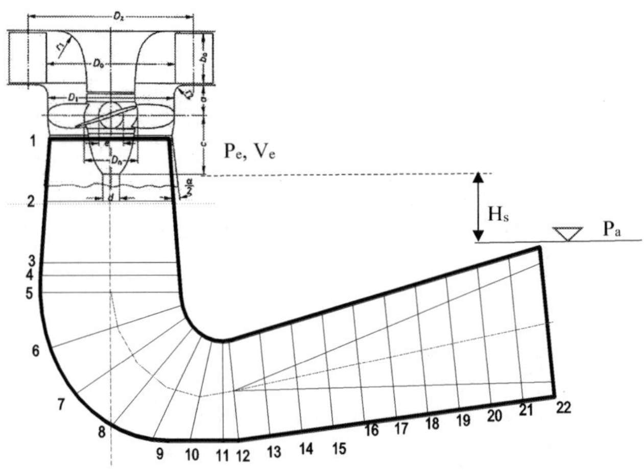

FIGURE 7.12 Turbine and draft tube.

From Equation 7.15, it follows that cavitation occurs when the pressure at the exit from the turbine is equal to the vapor pressure of the liquid at that point.

An important and useful parameter is a cavitation parameter (Thoma's cavitation coefficient), σ_p, defined as the ratio of exit velocity head and head of the machine. The Thoma's cavitation coefficient, which is named after the German Engineer Dietrich Thoma, is defined as follows.

$$\sigma_p = \frac{\dfrac{P_a - P_v}{\rho g} - \left(H_s - h_f\right)}{H} \tag{7.16}$$

To prevent cavitation, it is required that σ_p is greater than σ_c, where σ_c is the critical cavitation coefficient. Experimental measurements are used to determine σ_c, by plotting σ_p against efficiency.

Critical values of cavitation initiation have been correlated to specific speed of Francis turbines by the following empirical equation:

$$\sigma_p = 0.625\left(\frac{n_s}{100}\right)^2 \tag{7.17}$$

In Equation 7.17, specific speed is calculated using rpm, hp, and ft, where specific speed, N_s, is given by:

$$n_s = N\frac{\sqrt{P}}{H^{\frac{5}{4}}} \tag{7.18}$$

where

N is speed in rpm

P is power in horsepower

H is head in feet.

The critical cavitation value for Kaplan turbines is given by:

$$\sigma_p = 0.28 + \frac{1}{7.5}\left(\frac{n_s}{100}\right)^3 \tag{7.19}$$

7.3 METHODS OF IMPROVING CAVITATION PERFORMANCE OF PUMPS

If the primary goal is to improve cavitation properties at any cost regardless of the hydraulic performance of a pump, special design adjustments can be implemented. These adjustments are shown in Figures 7.13–7.15.

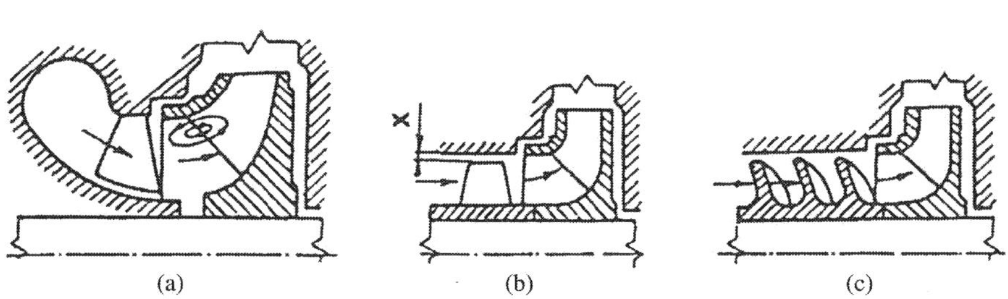

FIGURE 7.13 Improving cavitation properties of a pump using: (a) radial rib in suction; (b) booster; and (c) inducer.

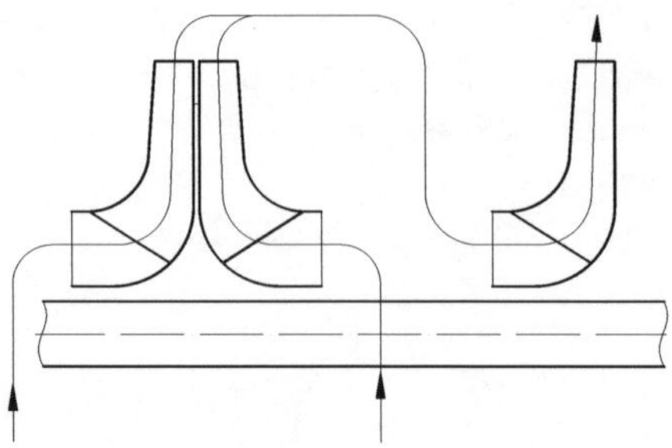

FIGURE 7.14 Impeller with extended entry.

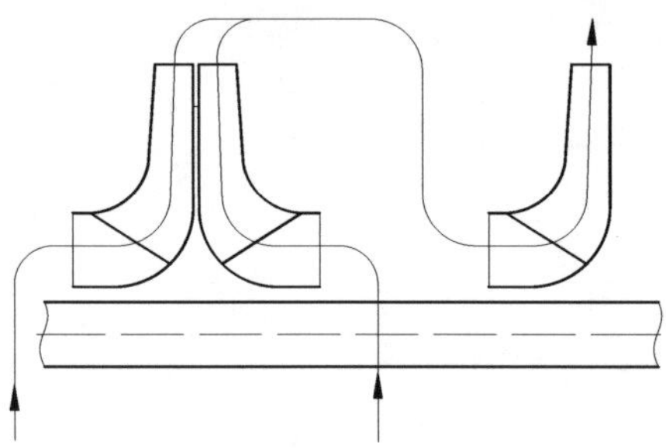

FIGURE 7.15 Double-entry impeller.

A pump is more resistant to cavitation, the lower its NPSH/σ_{crit}. For the same type of pump, the cavitation parameters can be reduced by design modifications, as shown in Figure 7.13. The radial rib in the pump suction area (Figure 7.13a) dampens the secondary flow in the impeller when $Q < Q_n$ and favorably affects the cavitation properties of the pump. A booster or inducer (Figure 7.13b and c) can be applied to increase the pressure at the inlet to the impeller. A pump inducer is a component that may be added to the axial intake part of a centrifugal pump rotor to elevate the inlet head to produce the needed pressure while preventing severe cavitation. The inducer increases the pressure by adding a circumferential component of the speed (Figure 7.16).

Cavitation properties of pumps can also be improved by an impeller with an extended inlet of multi-stage pumps (Figure 7.14) or with a double inlet (Figure 7.15). This is because a reduction in meridional velocities in the impeller inlet has a positive effect on the reduction of cavitation. Such design improvements can reduce cavitation by about 40%.

7.4 EXAMPLE PROBLEMS

Example 7.1

Water flows out of the tank through a continuously expanding pipe diffuser (Figure 7.17). Calculate h_1, the height at which the pressure in the narrowest section of

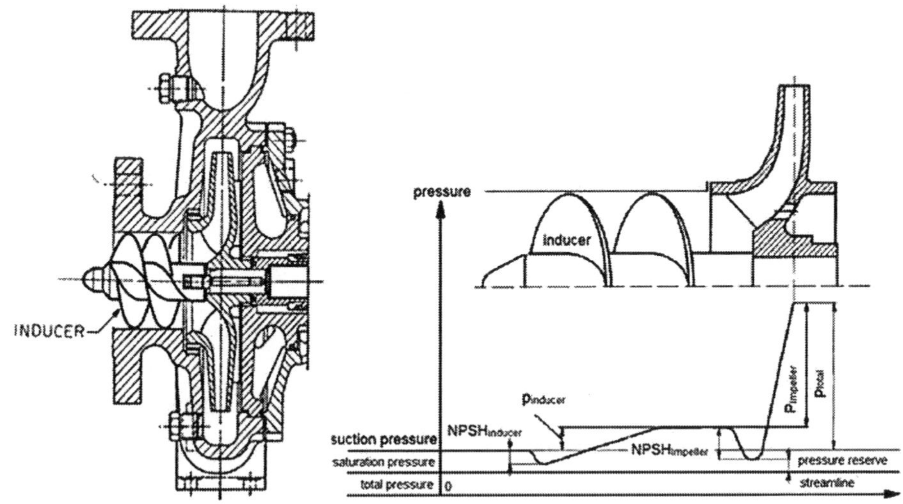

FIGURE 7.16 Pump with inducer.

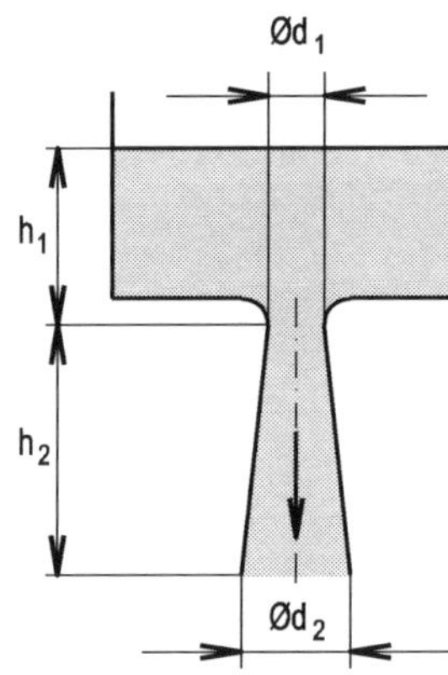

FIGURE 7.17 Water flow from a tank.

the diffuser equals the saturated steam pressure of the water P_v (water begins to evaporate – cavitation). Given: $d_1 = 100$ mm; $d_2 = 150$ mm; $h_2 = 1.15$ m; $P_a = 101.3$ kPa; $P_v = 4$ kPa.

Solution

Using Bernoulli's equation for points 0 and 2:

$$g \cdot (h_1 + h_2) = \frac{v_2^2}{2}$$

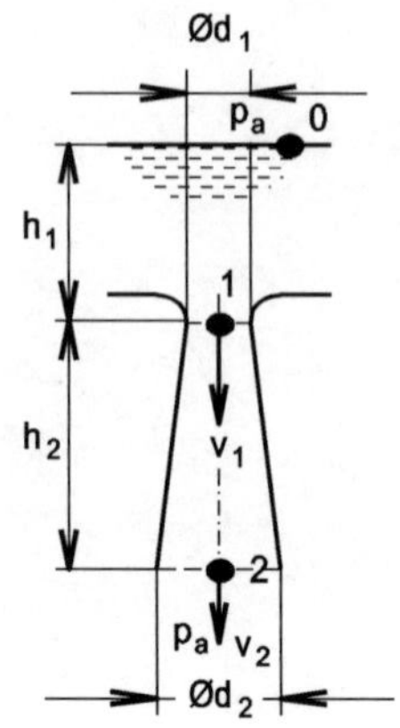

FIGURE 7.18 Water flow from a tank.

Using Bernoulli's equation for points 1 and 2:

$$\frac{V_1^2}{2}+\frac{p_v}{\rho}+g\cdot h_2=\frac{V_2^2}{2}+\frac{p_a}{\rho}$$

Equation of continuity: $v_1\cdot\frac{\pi\cdot d_1^2}{4}=v_2\cdot\frac{\pi\cdot d_2^2}{4}$

There are three unknowns in the above three equations: V_1, V_2, and h_1.

Solving we get: $v_2=\sqrt{2\cdot g\cdot\left(h_1+h_2\right)}$

$$v_1=\sqrt{2\cdot g\cdot\left(h_1+h_2\right)}\cdot\frac{d_2^2}{d_1^2}$$

$$h_1=\left(\frac{p_a-p_v}{\rho\cdot g}-h_2\cdot\frac{d_2^4}{d_1^4}\right)\cdot\frac{1}{\frac{d_2^4}{d_1^4}-1}=\left(\frac{101300-4000}{1000\cdot 9.81}-1.15\cdot\frac{150^4}{100^4}\right)\cdot\frac{1}{\frac{150^4}{100^4}-1}=1.008$$

Example 7.2

A liquid of density ρ flows out of a tank through the pipe (Figure 7.19). The rate at which the water level in the tank decreases is negligible. The pipe diameter is $Ød=0.12$ m; atmospheric pressure $P_a=101.33$ kPa; $z_2=2.5$ m. Calculate the velocity of the water leaving the pipe, the flow rate, and the pressure at the highest point of the pipe. Given are also: $z_1=0.8$ m; ρ = 1000 kg/m³, $P_v=6$ kPa. Determine the height, z_2, at which cavitation starts to occur.

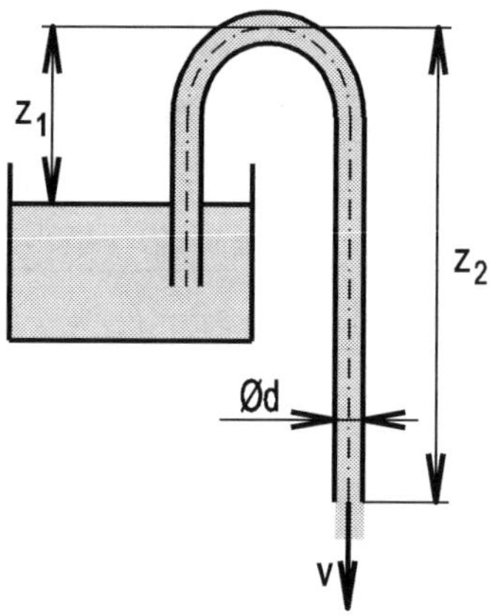

FIGURE 7.19 Water is siphoned through a pipe from a tank.

Solution (Figure 7.20)

Bernoulli's equation for points 0 and 2:

$$\frac{P_a}{\rho} + g \cdot (z_2 - z_1) = \frac{P_a}{\rho} + \frac{V_2^2}{2}$$

The outlet velocity is given by: $V_2 = \sqrt{2 \cdot g \cdot (z_2 + z_1)}$ (a)

$$V_2 = \sqrt{(2) \cdot (9.81) \cdot (2.5 - 0.8)} = 5.77 \mathrm{m\,s^{-1}}$$

Bernoulli's equation for points 0 and 1: $\dfrac{P_a}{\rho} = \dfrac{P_1}{\rho} + g \cdot z_1 + \dfrac{V_1^2}{2}$

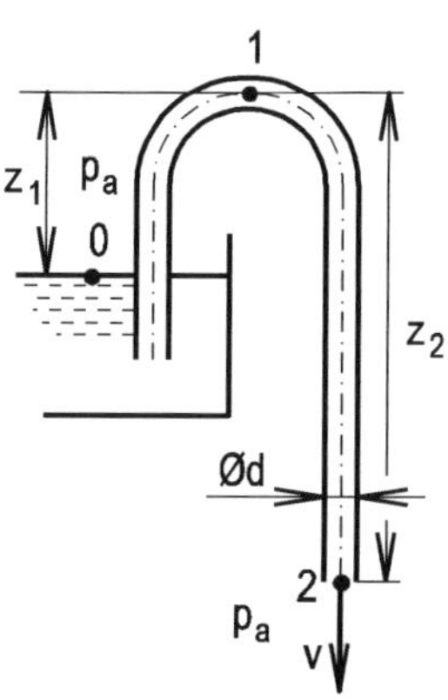

FIGURE 7.20 Water is siphoned through a pipe from a tank. Point 1 is the highest point.

P_1 can be obtained from this equation. Note that $V_1 = V_2$.

$$P_1 = P_a - \rho \cdot g \cdot z_1 - \rho \frac{V^2}{2} = 101330 - 9.81.1000.0.8 - 1000.\frac{5.77^2}{2} = 75.48\ kPa$$

Cavitation will start when $P_1 = P_v$. Writing the Bernoulli equation for points 1 and 2:

$\frac{P_v}{\rho} + g \cdot z_1 = \frac{P_a}{\rho} - g \cdot (z_2 - z_1) + \frac{V_2^2}{2}$, substituting V_2, which is given by equation (a) $V_2 = \sqrt{2 \cdot g \cdot (z_2 - z_1)}$ we get

$$z_2 = \frac{P_a - P_v}{\rho g} = \frac{(101.33 - 6)1000}{(9.81).(1000)} = 9.71\ m$$

Example 7.3

Water flows through a gravity pipe of diameter d, from a lake, as shown in Figure 7.21. The outlet end of the pipe is H meters below the level of the lake.

(a) What is the flow rate of the water?
(b) What can be the maximum value of H_2 to avoid cavitation at the highest point? Given: $l_1 = 100$ m, $l_2 = 200$ m, the total pipe length, $l = 300$ m, $d = 30$ mm, $P_v = 872$ Pa, $H = 9$ m, $P_a = 75.0$ kPa, $t = 15$°C, friction factor $f = 0.03$, minor loss coefficient, $\xi = 0.05$, $\rho = 1000$ kg/m^3.

Solution

Bernoulli's equation for the free surface and point 3 is given by:

$$\text{(a)} \quad \frac{P_a}{\rho} + g \cdot H = \frac{P_a}{\rho} + \frac{V^2}{2}\left(f\frac{l}{d} + \xi\right)$$

$$V = \sqrt{\frac{2 \cdot g \cdot H}{1 + f\frac{l}{d} + \xi}} = \sqrt{\frac{(2)(9.81)(9)}{1 + 300 + 0.05}} = 0.765\ m/s$$

$$Q = V\frac{\pi d^2}{4} = 0.765\frac{(3.14159)(0.03)^2}{4} = 0.54\ l/s$$

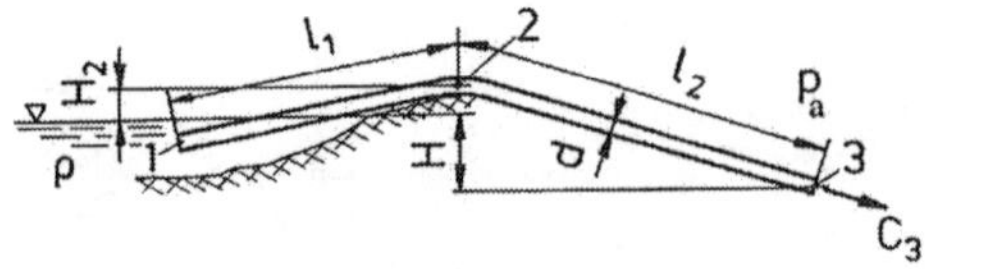

FIGURE 7.21 Gravity pipe.

(b) Bernoulli's equation between the free surface and the highest point (point 2)

$$\frac{P_a}{\rho} = g \cdot H_2 + \frac{P_v}{\rho} + \frac{V^2}{2} + \frac{V^2}{2}\left(f\frac{l_1}{d} + \xi\right)$$

$$H_2 = \frac{P_a - P_v - \frac{V^2}{2}\left(1 + f\frac{l_1}{d} + \xi\right)}{\rho g} = 4.54\ m$$

Example 7.4

What is the height of the suction pipe (H_{sg}) of the pump if sulfuric acid is pumped at a concentration of 60%? Density of sulfuric acid is 1500 kg/m^3 at temperature 20°C. The required pump flow rate is $Q = 0.025$ m^3/s and pump head $H = 80$ m. For these values, a single-stage pump, whose suction pipe has a diameter $D_s = 65$ mm and runs at a speed of $n = 2900$ rpm, was selected. The pump has cavitation properties defined by the manufacturer; $NPSH_R = 3.3$ m.

Determine the suction pipe height if:

(a) atmospheric pressure in the suction tank is given as $P_a = 1 \times 10^5$ Pa
(b) in the suction tank, the pressure is higher than atmospheric pressure by $P = 0.5 \times 10^5$ Pa
(c) in the closed suction tank, the pressure at the surface is equal to the saturated vapor pressure. $P_w = 380$ Pa for the given concentration and temperature. Assume the head losses in the suction pipe are estimated at 1.5 m.

Solution (Figure 7.22)

From Equation 7.4, $NPSH_A = \frac{P_a}{\rho g} - \frac{P_v}{\rho g} - H_s - h_f$, we get

$$H_{sg} = \frac{P_a}{\rho g} - \frac{P_v}{\rho g} - NPSH_A - h_f$$

To avoid cavitation $H_{sg} \le \frac{P_a}{\rho g} - \frac{P_v}{\rho g} - NPSH_A - h_f$

Case (a) $H_{sg} \le \frac{1\times 10^5 - 380}{(9.81)(1500)} - 1.5 - 3.3 = 1.96\ m$

Case (b) $H_{sg} \le \frac{(1+0.5)\times 10^5 - 380}{(9.81)(1500)} - 1.5 - 3.3 = 5.37\ m$

Case (c) $H_{sg} \le \frac{380 - 380}{(9.81)(1500)} - 1.5 - 3.3 = -4.8m$

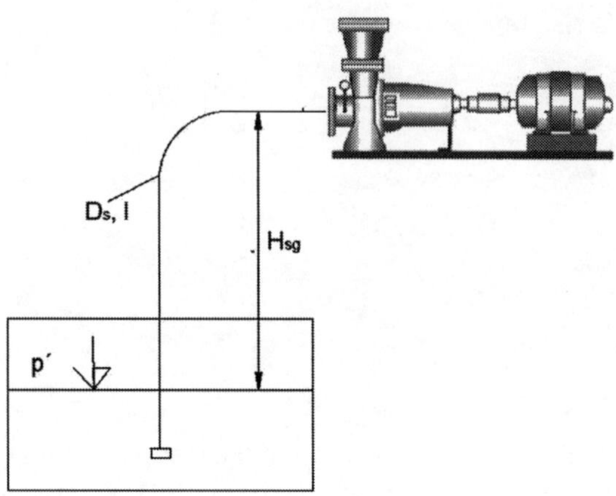

FIGURE 7.22 Determining the suction height of a pump.

In cases (a) and (b), the pump will be installed above the liquid surface level. In case (c), the pump must be installed below the surface. If the manufacturer did not give the value of $NPSH_R$ then we can calculate the critical cavitation coefficient and required NPSH from Thoma's cavitation coefficient.

$$NPSH_{crit} = \sigma_p H_{(1)}$$

Example 7.5

Calculate the suction height of the turbine shown, which has a head of 6 m. The specific speed is $n_s = 380$ rpm. The turbine will operate at an altitude of 900 m.

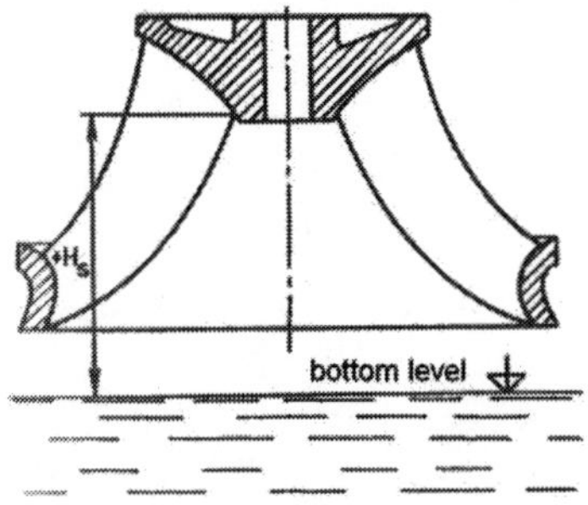

Solution

$\sigma_p = 0.33$ from Figure 7.10 or by interpolation from Table 7.2:

$$H_a = 10 - \frac{h}{100} = 10 - \frac{900}{900} = 9m$$

$$H_s = H_a - \sigma_p H = 9 - (0.33)(6) = 7.02m$$

7.5 EXERCISE PROBLEMS

Problem 7.1

Determine the permissible suction head of a pump for the system shown on the right-hand side:

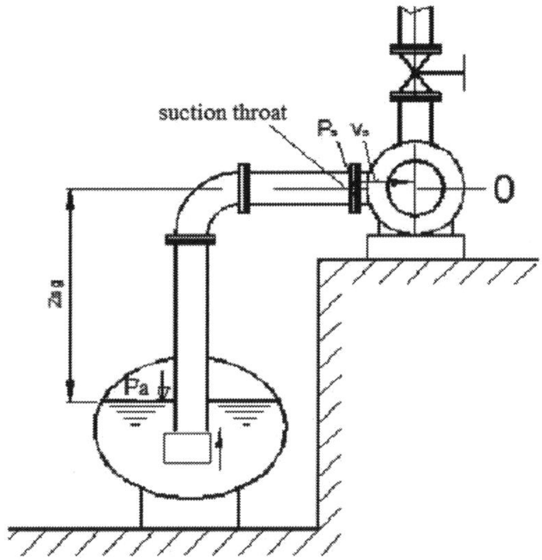

Given: $z_{sg} = 2$ m, $Q = 120$ l/min $= 2$ l/s, $P_a / \rho g = 10.33$ m, $d_s = 50$ mm, $\text{NPSH}_a = 2.5$ m, saturated water vapor at 20°C, $P_w / \rho g = 0.52$ m.

The length of suction pipe $l_s = 20$ m, friction factor coefficient of length losses $f = 0.02$, minor loss coefficients of the elbow, the strainer, and the valve are $\xi_k = 3$, $\xi_s = 3$, $\xi_v = 6$, respectively.

Problem 7.2

The pump's available NSPH is 3 m. The flow rate is $Q = 2400$ l/s and suction pipe diameter $d_s = 200$ mm. Friction coefficient $f = 0.02$, minor loss coefficients of the elbow, strainer, and valve are $\xi_k = 0.25$, $\xi_s = 3$, $\xi_v = 6$, respectively. Determine whether cavitation can occur in this pump.

Problem 7.3

Calculate the suction height of the turbine shown, which has a head of 30 m. The specific speed is $n_s = 380$ rpm. The turbine will operate at an altitude of 900 m.

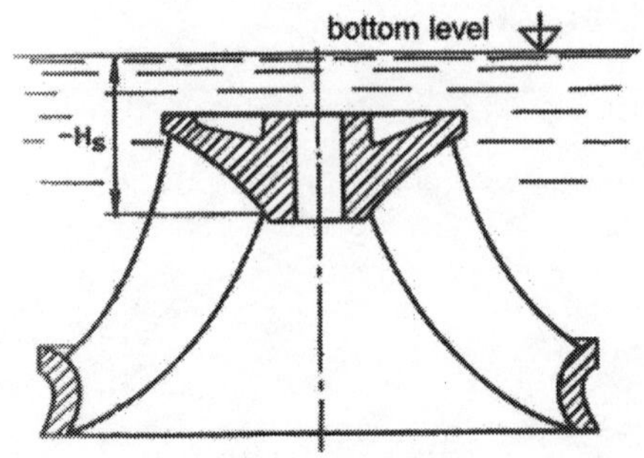

Problem 7.4

Calculate the suction height of the turbine shown, which has a head of 30 m. The specific speed is $n_s = 600$ rpm. The turbine will operate at an altitude of 900 m.

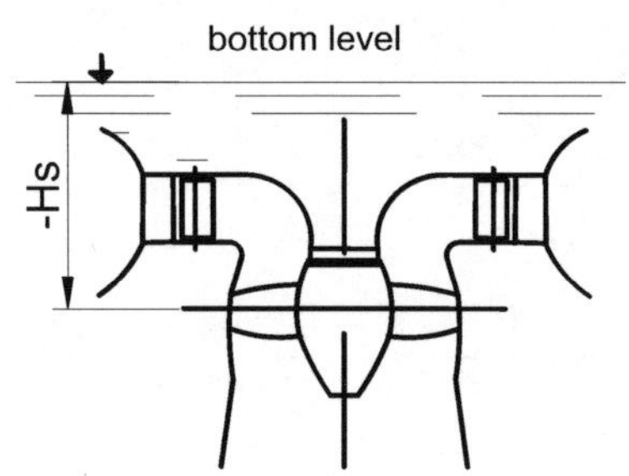

Problem 7.5

A centrifugal pump transports water from an open container. The suction flange E is connected by a suction pipe to the tank, where there is a suction head, H_s. Water flows through the suction pipe at a velocity of v_s. The losses in the suction pipe are estimated as H_{sloss}. Given: $H_s = 5.5$ m, $H_{sloss} = 4.3$ m, $P_v = 2643$ Pa, NPSH = 5.5 m, Pa = 93 kPa, $v_s = 4.7$ m/s, ρ = 997 kg/m³. Will this design condition prevent cavitation? If not, determine the permissible impeller inlet pressure P_E, so that cavitation does not occur.

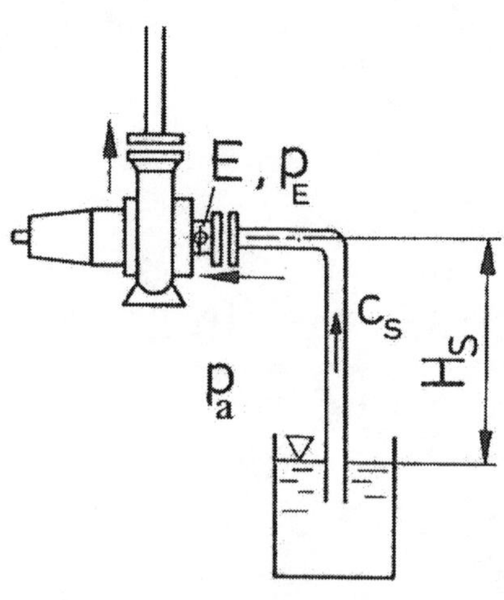

Problem 7.6

A draft tube with efficiency η_D is connected to the water turbine.

(a) Determine the static pressure P_s at the inlet to the draft tube at point S if $H_s = 3.5$m.

(b) Determine H_s if the pressure at point S is the saturated vapor pressure at a given temperature.

Given: $c_s = 9$ m/s, $c_A = 3$ m/s, $H_s = 3.5$ m, $P_a = 95$ kPa, $\eta_D = 0.90$, $P_v = 1704$ Pa at $t = 15°C$, $\rho = 1000$ kg/m³.

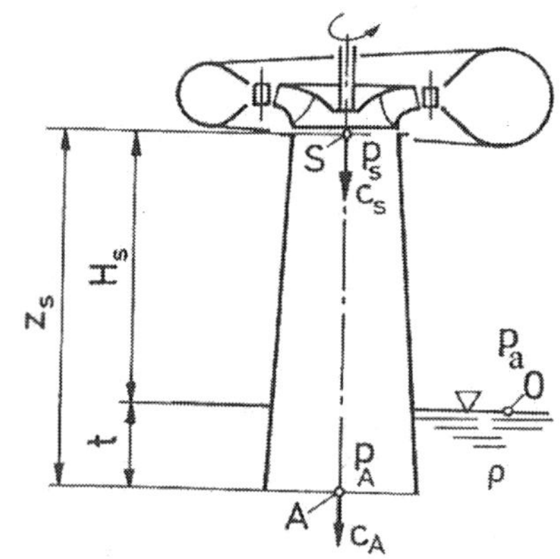

Problem 7.7

Pressure at the exit from a Kaplan turbine is measured with a mercury U manometer, of which the reading is given by Δh. For the following given values, $P_a = 96{,}000$ Pa, $c_1 = 6.2$ m/s, $c_A = 1.3$ m/s, $z = 1.2$ m, $y = 1.05$ m, $H_s = 2.2$ m, pressure loss from points 1 and A is given by: $\Delta p_{sloss} = 0.15.\rho.\frac{c_1^2}{2}$, $\rho = 1000$ kg/m³

is density of water and density of mercury, A= 1000 kg/m³, determine: (a) pressure P_1 and Δh.

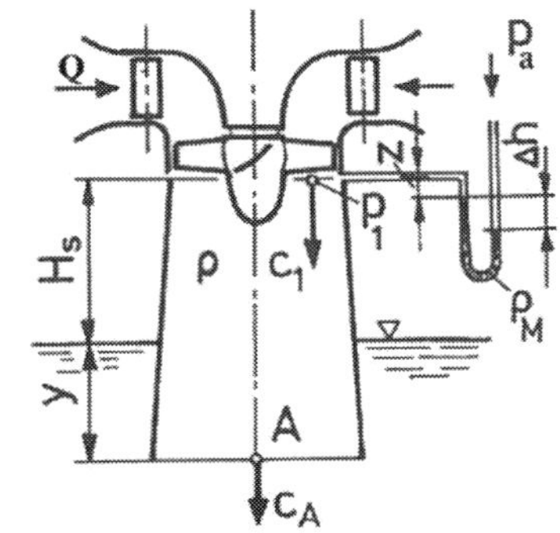

Problem 7.8

Determine the height of of the suction pipe of a pump which transports H_2SO_4 from a container. H_2SO_4 has 60% concentration; $\rho = 1500$ kg/m^3 at 20°C. The flow rate and the head of the pump are $Q = 0.025$ m^3/s and $H = 80$ m, respectively. A single-stage pump is used for this purpose. The diameter of the suction pipe $D_s = 65$ mm and rotates at $N = 2900$ rpm. $NPSH_A = 3.3$ m and H_{sloss} is 1.5 m. P_v for the given temperature and concentration is 380 Pa.

Consider the following conditions:

(a) Pressure in the container is atmospheric, $P_a = 10^5$Pa.
(b) The container is closed and pressure at the surface is 0.5×10^5 Pa.
(c) The pressure in the closed container is equal to the saturated vapor pressure.

7.6 BIBLIOGRAPHY

Blaha, J. and Brada, K. *Pumping Technology Manual*. Prague: ČVUT, 1997.

Brennen, C. E. *Cavitation and Bubble Dynamics*. New York: Oxford University Press, 2013.

Duan, C. G., Karelin, V. I. and others, *Abrasive Erosion and Corrosion of Hydraulic Machinery: Series on Hydraulic Machinery*. London: Imperial College Press, 2003.

Franc, J.-M. and Michel, J.-P. *Fundamentals of Cavitation*. Netherlands: Springer, 2005.

Gülich, J. F. *Centrifugal Pumps*. Berlin: Springer, 2014.

Karassik, I. J., Krutzsch, W. C., Fraser, W. H. and Messina, J. P. *Pump Handbook*. New York: McGraw Hill, 1976.

Kozák, J. Cavitation Induced by Rotation of Liquid. Brno: VUT Brno, 2015.

Michel, J. M. "Introduction to cavitation and supercavitation," in *RTO AVT Lecture Series on "Super-cavitating Flows"*, 2001.

Noskievič, J. *Cavitation in Hydraulic Machines and Equipment (in Czech)*. Prague: SNTL, 2005.

Noskievič, J. *Cavitation (in Czech)*. Prague: Academia, 1969.

Nourbakhsh, H. B. P. A., Jaumotte, A., Hirsch, C. *Turbopumps and Pumping Systems*. Berlin: Springer, 2008.

Paciga, G. M. and Strýček, O. *Pumping Techniques (in Slovak)*. Bratislava, Slovakia: ALFA, 1984.

Strýček, O. *Hydrodynamic Pumps*. Bratislava, Slovakia: STU Bratislava, 1994.

Van den Braembussche, R. *Introduction to Cavitation and Supercavitation*. Neuilly-sur-Seine: RTO/NATO, 2002.

Varchola, M. and Hlobocan, P. "Hydraulic interaction between an impeller and axial diffuser of a mixed-flow pump (in Slovak)," in *Current Trends in Development of Pumping Machinery*, Applied Mechanics and Materials (Vol. 630), pp. 35–42, 2013. Available at: www.scientific.net/AMM.630.35.

Varchola, M. and Hlobocan, P. *Hydraulic Design of Centrifugal Pumps*. Bratislava, Slovakia: STU Bratislava, 2016.

Willi Bohl, W. E. *Fluid Flow Machines 1. Structure and Mode of Operation (Kamprath Series) (in German)*. Würzburg: Vogel Communications Group.

8 Testing Hydraulic Machines

8.1 TESTING FACILITIES

Hydraulic machine tests are carried out for the following purposes:

1. to verify performance and suitability for service
2. to identify the causes of failures
3. to develop new or better hydraulic components.

Following the completion of the assembly, the pump manufacturer conducts a pump performance test to ensure that the pump meets the required specifications as stated in the pump datasheet and other purchase documentation. After the performance test, the net positive suction head (NPSH) test, mechanical running test, and final inspection are performed. Vibration testing is done during the performance test as well as the mechanical running test. Pump testing can be done in the field or in a laboratory setting.

Acceptance tests are generally performed in accordance with industry standards, such as ISO 9906, ASME PTC 8.2 Centrifugal Pumps Power Test Code, Hydraulic Institute (HI) Standards, Section: Centrifugal pumps, test standards (Cleveland, USA).

8.2 TESTING SETUPS

A typical closed-loop pump testing setup is shown in Figure 8.1. Before the test, the system must be primed, which means that all air in the system must be vented from the system's highest point. The noncavitation test is often performed with the throttle valve completely open. The valve is then closed to lower the flow rate. The rotation speed may vary during the tests. The flow rate, head (pressure), and shaft power should be adjusted from observed values at test speed to rated speed before plotting. Similarity laws are used to make the adjustments.

$$\frac{Q_1}{Q_2}=\frac{N_1}{N_2} \qquad \frac{\Delta p_1}{\Delta p_2}=\left(\frac{N_1}{N_2}\right)^2 \qquad \frac{P_1}{P_2}=\left(\frac{N_1}{N_2}\right)^3 \tag{8.1}$$

where Q is flow rate, N is rotational speed, Δp is pressure, and P is power. Subscripts 2 and 1 refer to test and rated speed, respectively. During the test, the values of speed, discharge and suction pressures, temperature, flow rate, and power should be recorded concurrently at each test point. Each quantity is recorded several times in a brief

DOI: 10.1201/9781003007142-8

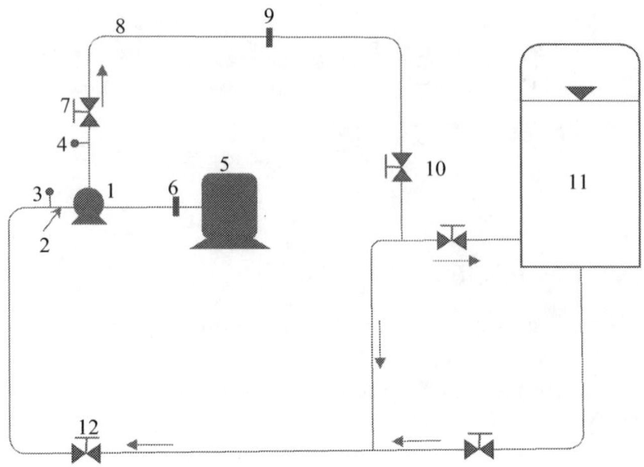

FIGURE 8.1 Typical closed-loop pump testing setup: (1) pump; (2) suction pipe; (3) suction pressure gauge; (4) discharge pressure gauge; (5) electric motor; (6) coupling; (7) valve; (8) discharge pipe; (9) flow meter; (10) valve; (11) tank; (12) valve.

timeframe when utilizing an automated data-recording system. Such measurements are averaged, and the scatter (standard deviation) is calculated and compared to the permitted variations.

The number of test points required to obtain the head curve is determined by the curve's steepness. Steep curves require a few points, whereas flat or unstable curves require more points to ensure that an instability is detected. If noticeable instability is found, one test from high to low flow and another from low to high flow may be performed.

8.3 INSTRUMENTS AND MEASUREMENTS

The main parameters to be measured are suction and discharge pressures, volumetric flow rate, rotating speed of the impeller, input shaft torque or power, and fluid temperature. Acceptance test measuring instruments must fulfill rigorous precision standards and must be calibrated both officially and in-house according to the relevant standard. The test class of the acceptance standard that is utilized defines these accuracy criteria.

8.3.1 Pressure Measurements

Pressure can be measured using manometers (water or mercury), pressure gauges, or pressure transducers.

8.3.2 Flow Rate Measurements

Flow rate can be measured by standard orifice plates, standard Venturi nozzle, or Venturi tube.

8.3.3 Speed Measurements

Speed measurements are frequently conducted with digital counters that count discrete pulses with each rotation. The impeller speed can be measured with tachometer, stroboscope, or photocell.

8.3.4 Shaft Power Measurements

The shaft power can be determined with transmission dynamometer or torsion dynamometer. Note that these measurement devices are given as an example, but many other devices are being developed and used.

8.3.5 Pump Performance Characteristics Determination

The measured raw data can be used to plot performance characteristics. The performance curves of interest are pump head (pressure), power, and efficiency curves. The $NPSH_R$ curve is also of interest. The NPSH test will be discussed separately.

The three most commonly used graphical representations of pump performance are:

1. change in total head produced by the pump, H_t
2. power input to the pump, P_m
3. pump efficiency, η.

Total head

The change in total head produced as a result of the work done by the pump can be calculated as:

$$H_t = \textit{change in static head} + \textit{change in velocity head} + \textit{change in elevation}$$

$$H_t = H_s + H_v + H_e \tag{8.2}$$

where

$$H_s = \textit{Change in static head} = \frac{P_{out} - P_{in}}{\rho g} \tag{8.3}$$

where

P_{in} = *fluid pressure at inlet in Pa*
P_{out} = *fluid pressure at outlet in Pa*
H_s = *change in static head in meters*

$$H_v = Change\,in\,velocity\,head = \frac{\left(V_{out}^2 - V_{in}^2\right)}{2g} \quad (8.4)$$

where
V_{in} = fluid velocity at inlet in m/s
V_{out} = fluid velocity at outlet in m/s
H_e = Change in elevation in m = vertical distance between inlet and outlet

Power input

The mechanical power input to the pump is calculated as:

$$P_m = 2\pi n\tau \quad (8.5)$$

where
n = rotational speed of the pump in revloutions per second
τ = shaft torque in Nm

Note: $P_m = \tau\omega$, where ω is in s^{-1}.

Pump efficiency

The efficiency of a pump in percent may be calculated as:

$$\eta = \frac{P_h}{P_m} \times 100 \quad (8.6)$$

where

$$P_h = hydraulic\ power\ imparted\ to\ fluid$$
$$P_h = H_t Q\rho g \quad (8.7)$$

where
Q = volume flow rate in m^3/s
P_m = mechanical power abosrbed by pump
$P_m = 2\pi n\tau$

Each of these parameters is measured at constant pump speed, and is plotted against the volume flow rate, Q, through the pump. If the speed is not constant, we use the similarity law equations given in Equation 8.1.

8.3.6 NPSH Tests

The goal of the NPSH test is to ensure that the pump's $NPSH_R$ meets the guarantee requirements. According to the HI, this test is only concerned with data relating to the

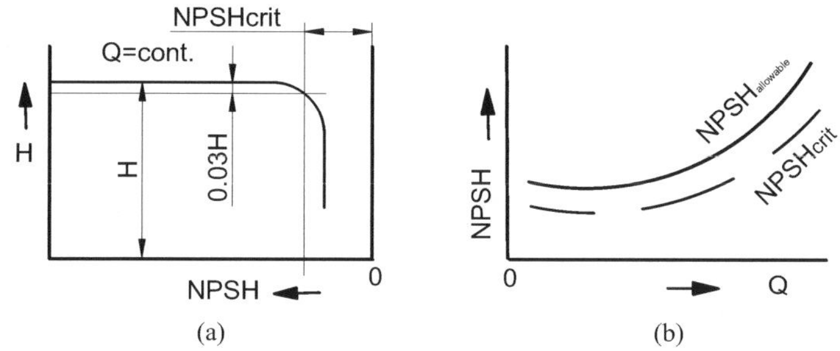

FIGURE 8.2 Cavitation characteristics: (a) critical cavitation depression; (b) cavitation characteristics.

pump's hydraulic performance (head, flow, and power fluctuations), not with other cavitation-related effects (e.g., noise, vibration, and erosion). At a given rate of flow, cavitation effects can be observed as a reduction in head or power. Unless otherwise indicated, $NPSH_R$ will be calculated using a 3% reduction in head (standard industry practice) and designated as $NPSH_3$ (Figure 8.2).

In multi-stage pumps, the head drop refers to the first stage's head, which should be measured if available. A head drop of more than 3% may be agreed upon for very low head pumps. Cavitation tests are usually performed using deaerated clear water. Cavitation experiments in water are insufficient to predict the pump's performance in liquids other than clear water.

8.3.6.1 NPSH Test Process

A typical NPSH testing setup is shown in Figure 8.3. The test procedure is that the pressure in the suction tank is gradually reduced while maintaining a constant flow rate, Q.

Maintaining speed and flow rate constant while reducing the available NPSH in the test loop is the suggested technique for getting the most accurate cavitation test results possible. The change in head is observed as shown in Figure 8.2. The critical value of $NPSH_{crit}$ that corresponds to a 3% reduction in head is determined. It means the suction pressure is gradually reduced until 3% reduction in pump head is achieved. Then the test is stopped and $NPSH_R$ is recorded (Figure 8.3). Then the flow rate is changed, and the procedure is repeated. From the measurement with multiple flow rates, we obtain the cavitation characteristic shown in Figure 8.3. $NPSH_R$ is the required pressure in the impeller eye, and $NPSH_A$ is the available suction pressure; to avoid cavitation, the suction pressure should always be greater than the vapor pressure.

Another way of performing an NPSH test utilizes the procedure outlined below. In this test, instead of doing the above-mentioned suction tests at a constant flow rate, the $NPSH_R$ can be determined by changing the flow rate as shown in Figure 8.4.

In this test, the suction valve's position is maintained while the discharge valve is gradually opened. The flow rate increases along the characteristic for cavitation-free operation (curve "a" in Figure 8.4), until cavitation causes the head to collapse (points AB). The flow rate increases as the discharge valve is opened further, and

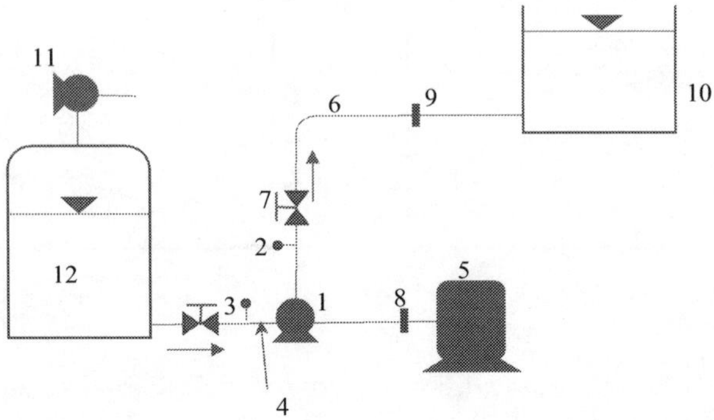

FIGURE 8.3 Pump test circuit suitable for net positive suction head (NPSH) testing: (1) ump; (2) discharge pressure gauge; (3) suction pressure gauge; (4) suction pipe; (5) electric motor; (6) discharge pipe; (7) valve; (8) coupling; (9) flow meter; (10) discharge tank; (11) vacuum pump; (12) suction tank.

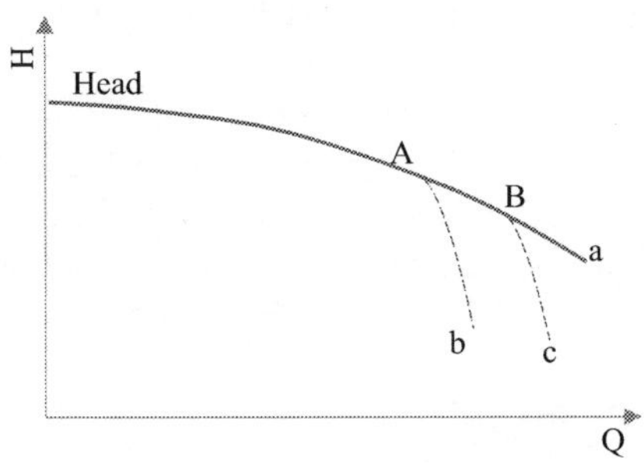

FIGURE 8.4 Net positive suction head (NPSH) test at constant suction valve position.

the head drops owing to cavitation. This test procedure is less accurate than the suction tests.

8.4 EXAMPLE PROBLEMS

Example 8.1

A small centrifugal pump was tested in a laboratory setting using a closed-loop test rig shown in Figure 8.5. As shown in the figure, water is pumped (using pump 1) from the tank. Valve 1 is fully open and valve 2 is fully closed initially. Flow rate is regulated by gradually opening valve 1. The following parameters (shown in Table 8.1) were measured. Establish the performance curve for this pump.

Solution

Equation.8.2 to Equation 8.5 were used to determine the mechanical power (P_m), hydraulic power (P_h), theoretical head (H_t), and efficiency (η) (Figure 8.6).

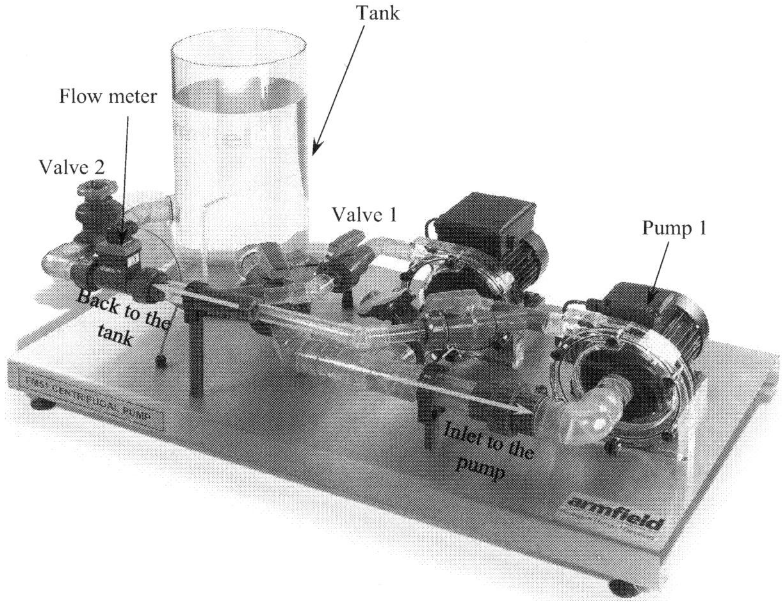

FIGURE 8.5 Experimental setup.

The calculated values are given in blue in Table 8.2. The following simple code in EES was used for the calculation:

$$\rho = \rho(\text{Water},\ \text{T} = 16,\ \text{P} = 101.325)$$
$$g = 9.806\ \left[\text{m/s}^2\right]$$
$$p = \pi$$
$$n = (1440/60)$$
$$H_s = \frac{(P_{out} - P_{in}) \cdot 1000}{\rho \cdot g}$$
$$H_v = \frac{(V_{out})^2 - (V_{in})^2}{2 \cdot g}$$
$$H_t = H_s + H_v + H_e$$
$$P_m = 2 \cdot p \cdot n \cdot \tau$$
$$P_h = H_t \cdot Q \cdot \rho \cdot g \cdot 0.001$$
$$\eta = (P_h / P_m) \cdot 100$$

Performance testing for extremely large pumps can be extremely costly. Instead of testing the full-size pump, model acceptability tests can be performed. Usually, it is recommended that a scaling ratio of no more than ten be used. It is important that the

TABLE 8.1
Measured Parameters

Sample number	Pump Setting S [%]	Pump speed n [rpm]	Water Temperature T [°C]	Inlet Pressure P_0 [kPa]	Pump 1 pressure P_1 [kPa]	Motor 1 Torque t [Nm]	Flow fate Q [l/s]	Operating Mode	Density of Water rho [kg/m³]	Pump 1 Inlet Velocity [m/s]	Pump 1 Outlet Velocity [m/s]
1	80	1440	16.4	3.8	60.8	0.67	0.000	Single	999	0.00	0.00
2	80	1440	16.5	3.6	57.5	0.68	0.077	Single	999	0.18	0.32
3	80	1440	16.5	3.6	56.8	0.71	0.154	Single	999	0.35	0.64
4	80	1440	16.6	3.5	55.8	0.80	0.346	Single	999	0.80	1.44
5	80	1440	16.6	3.5	53.1	0.89	0.596	Single	999	1.37	2.48
6	80	1440	16.6	3.2	48.8	0.97	0.846	Single	999	1.95	3.52
7	80	1440	16.6	3.2	45.0	1.07	1.019	Single	999	2.35	4.24
8	80	1440	16.6	2.7	40.7	1.13	1.134	Single	999	2.62	4.72
9	80	1440	16.6	2.7	38.3	1.14	1.211	Single	999	2.79	5.04
10	80	1440	16.6	2.9	37.3	1.19	1.269	Single	999	2.93	5.28
11	80	1440	16.7	2.4	36.3	1.20	1.288	Single	999	2.97	5.36
12	80	1440	16.8	3.0	36.2	1.19	1.308	Single	999	3.01	5.44
13	80	1440	16.8	2.8	34.6	1.23	1.327	Single	999	3.06	5.52
14	80	1440	16.8	2.3	34.7	1.22	1.365	Single	999	3.15	5.68
15	80	1440	16.8	2.5	32.8	1.23	1.365	Single	999	3.15	5.68

TABLE 8.2
Table of Measured and Calculated Value

Sample number	Q [l/s]	V_{in} [m/s]	V_{out} [m/s]	τ [Nm]	P_{in} [kPa]	P_{out} [kPa]	P_m [W]	P_h [W]	H_t [m]	η [%]
1	0	0	0	0.67	3.8	60.8	101	0	5.894	0
2	0.077	0.18	0.32	0.68	3.6	57.5	102.5	4.21	5.581	4.105
3	0.154	0	0.64	0.71	3.6	56.8	107.1	8.337	5.527	7.787
4	0.346	0.8	1.44	0.8	3.5	55.8	120.6	18.6	5.487	15.42
5	0.596	1.37	2.48	0.89	3.5	53.1	134.2	31.27	5.356	23.3
6	0.846	1.95	3.52	0.97	3.2	48.8	146.3	42.83	5.168	29.28
7	1.019	2.35	4.24	1.07	3.2	45	161.4	49.68	4.977	30.79
8	1.134	2.62	4.72	1.13	2.7	40.7	170.4	52.66	4.74	30.9
9	1.211	2.79	5.04	1.14	2.7	38.3	171.9	54.66	4.608	31.79
10	1.269	2.93	5.28	1.19	2.9	37.3	179.4	56.81	4.571	31.66
11	1.288	2.97	5.36	1.2	2.4	36.3	181	57.42	4.551	31.73
12	1.308	3.01	5.44	1.19	3	36.2	179.4	57.8	4.511	32.21
13	1.327	3.06	5.52	1.23	2.8	34.6	185.5	57.16	4.398	30.82
14	1.365	3.15	5.68	1.22	2.3	34.7	184	60.46	4.522	32.86
15	1.365	3.15	5.68	1.23	2.5	32.8	185.5	57.59	4.307	31.05

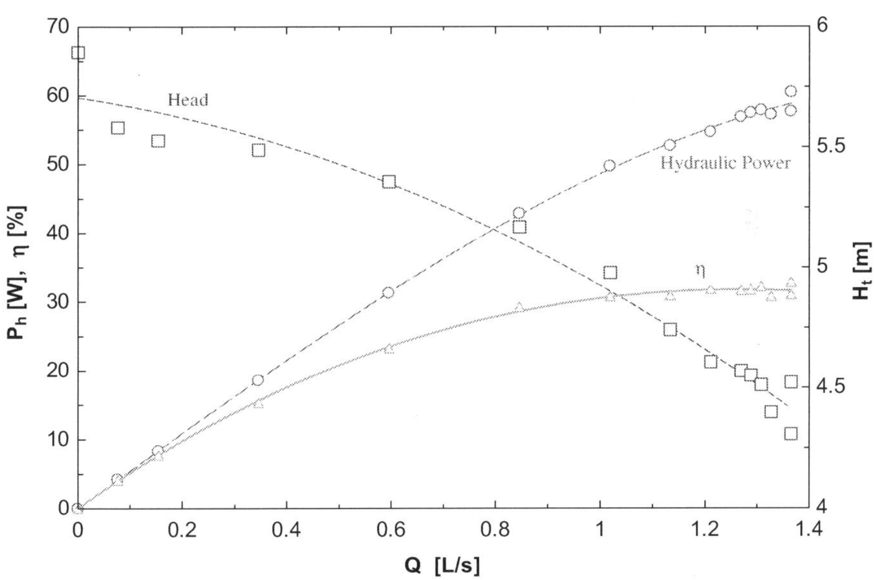

FIGURE 8.6 Performance characteristics of the tested pump.

model is manufactured precisely with allowable tolerances to ensure that the model performance tests are indicative of the performance of the full-scale prototype.

8.5 FIELD TESTS

Field tests are occasionally conducted to evaluate performance, particularly efficiency, under real-world conditions, such as full speed and temperature. However, determining efficiency with sufficient precision to justify the considerable work and cost required is challenging. Field tests are increasingly being used to determine the source of problems and to assess the efficacy of proposed solutions.

8.6 BIBLIOGRAPHY

Benra, F. K. "Measurement of the characteristics of a centrifugal pump," *Pract. Course Turbomach.*, 2013. Available: www.uni-due.de/sm/Downloads/Praktika/Centrifugal_Pump.pdf.

Gülich, J. F. "Operation of centrifugal pumps," in *Centrifugal Pumps*. Berlin: Springer, 2020.

Peng, W. W. *Fundamentals of Turbomachinery*. Hoboken, NJ: John Wiley, 2007.

Ramadevi, R. "Net positive suction head analysis for testing the condition of a centrifugal pump," *Indian J. Sci. Technol.*, 8 (10), 934–939, 2015.

9 CFD Analysis in Turbomachinery

9.1 INTRODUCTION

Real fluid flows are described by partial differential equations. These partial differential equations, in the general case, cannot be solved analytically. These equations can be approximated by numerical techniques by discretizing a complicated flow domain into several small cells. Numerical fluid flow calculations known as computational fluid dynamics (CFD) have become a distinct branch of fluid dynamics due to their wide applications in the general thermo-fluids area. Therefore, CFD is a part of fluid mechanics, which uses methods and tools of numerical mathematics and computer science. The field of CFD focuses on mathematical modeling of physical processes, including mainly fluid flow and heat transfer. For this purpose, numerical calculation methods are applied, while the calculations themselves are performed using computers. CFD is applied to many areas of thermo-fluids, including phase changes (evaporation, condensation, solidification, melting), cavitation, solid-particle interactions with fluids, chemical reactions (combustion processes), vibro-acoustics, and magnetohydrodynamics.

The standard formulation of the CFD simulation of a hydrodynamic machine usually involves calculation of the desired hydraulic parameters of the machine for given geometry and boundary conditions. The use of CFD simulation in the hydraulic design of a hydrodynamic machine involves the gradual modification of the geometry in order to achieve specific hydraulic parameters at pre-specified boundary conditions. After each modification of the geometry, a CFD simulation is performed, to answer the question whether the performed modification ensured the requirements of the hydraulic design, i.e., whether the geometry meets the requirements or not. Such "tuning" of the geometry to meet the requirements of the hydraulic design can be done "manually," with an important role played by the experience and knowledge of the CFD software (including the use of previous research to analyze the influence of geometry on the hydraulic properties of hydrodynamic machines). The experience of the person who performs the simulation work is important. Based on her/his own discretion and on the basis of detailed postprocessing (specific CFD simulations) s/he decides how to modify the geometry of the hydrodynamic machine in the next step so as to achieve the required parameters resulting from the hydraulic design assignment. In addition to the above "manual" geometry tuning, it is possible to apply "automatic" geometry optimization of a hydrodynamic machine. In the first step, the initial primary geometry is selected, which is then subjected to CFD simulation. Subsequently, based on a specific algorithm, selected geometric parameters are automatically modified, while each geometric alternative is subjected to CFD

DOI: 10.1201/9781003007142-9

analysis. The process is repeated until the required parameters of the hydraulic design are satisfied. A separate category of design procedures consists of algorithms for geometry design based on the inverse design. In the first step, the required flow field is defined, and, by gradual modifications of the geometry, a solution is sought that suits the required flow field.

CFD simulation can be performed either on the entire interior of the machine or only in some parts. In the case of CFD simulation of the entire interior of a hydrodynamic machine, interfaces are applied between the rotating and stationary parts of the machine, enabling mathematical solution to the rotating and stationary parts in one simulation. In the case of a CFD simulation of a separate impeller or diffuser, the mutual interaction of the impeller and the diffuser is not taken into account in the calculation. This limits the accuracy of the solution and the results approach reality only in a narrow area around the best-efficiency point (BEP).

9.2 CFD METHODOLOGY

9.2.1 Mathematical–Physical Model of Flow in a Hydrodynamic Machine

This section attempts to explain the basic equations describing single-phase flow in a hydrodynamic machine without considering phase changes and particulate flow. In general, in the field of CFD, the following assumptions are made. The flowing fluid is considered a continuum. The following conservation laws are assumed to be valid:

- mass
- momentum
- energy.

The continuity equation follows from the law of conservation of mass. The conservation of momentum, inertia, mass, and surface forces acting on a flowing fluid is expressed by either the Euler (nonviscous) or Navier–Stokes equations (viscous flow). Due to the detailed simulation of flow in hydrodynamic machines, three-dimensional (3D) viscous flow is commonly used. Due to the high Reynolds number, the flow in hydrodynamic machines is turbulent. The energy equation follows from the law of conservation of energy. Depending on whether the subject of CFD analysis is performed for a stationary or dynamic phenomenon, it is necessary to derive the equations describing the flow for either steady or unsteady flow.

Navier–Stokes equations for a real fluid can be expressed in the form of Equation (9.1). In a rotating coordinate system (impeller space), the left side of Equation (9.1) must be supplemented by a term which takes into account the Coriolis and centrifugal forces, Equation (9.2). Instead of absolute velocities, then relative velocities are used in the Navier–Stokes equations.

$$\frac{\partial}{\partial t}(\rho\mathbf{u})+\nabla\cdot(\rho\mathbf{u}\cdot\mathbf{u})=-\nabla p+\nabla\cdot\boldsymbol{\tau}+\mathbf{f}_v \tag{9.1}$$

$$\rho\left(2\boldsymbol{\omega}\times\boldsymbol{u}+\boldsymbol{\omega}\times\boldsymbol{\omega}\times\boldsymbol{r}\right) \tag{9.2}$$

where

p is pressure (Pa)
τ is shear stress tensor (Pa)
f_v is unit volume force (N)
ω is angular velocity vector (rad/s)
r is position vector (m).

The continuity equation (for both compressible and incompressible flow) takes the form of Equation (9.3).

$$\frac{\partial}{\partial t}\rho+\nabla\cdot\left(\rho\mathbf{u}\right)=0 \tag{9.3}$$

where

t is time (s)
ρ is density (kg/m^3)
u is velocity vector (for a stationary coordinate system it is an absolute velocity and for a rotating system it is a relative velocity) (m/s).

In steady flow the terms containing time derivatives in Equations (9.1–9.3) are equal to zero. For incompressible flow, the density is constant, so the equations can be further simplified. When expressing variables with components (three velocity components) or nine stress components, it is advantageous to use a specially abbreviated notation with well-defined rules, which are known as Einstein summation convention. According this convention, only one member can express all three components of velocity or stress components. As an example, below is the continuity equation for a steady, incompressible flow.

$$\frac{\partial u_1}{\partial x_1}+\frac{\partial u_2}{\partial x_2}+\frac{\partial u_3}{\partial x_3}=0 \quad \text{or} \quad \sum_{j=1}^{n}\frac{\partial u_j}{\partial x_j}=\frac{\partial u_j}{\partial x_j}=0 \tag{9.4}$$

The Navier–Stokes equations are given as follows (Equations 9.5 and 9.6):

$$\frac{\partial u_i}{\partial t}+\sum_{j=1}^{n}\frac{\partial\left(u_i u_j\right)}{\partial x_j}=F_i-\frac{1}{\rho}\frac{\partial p}{\partial x_i}+\nu\sum_{j=1}^{n}\frac{\partial^2 u_i}{\partial x_j^2}+e_z \tag{9.5}$$

$$\frac{\partial u_i}{\partial t}+\frac{\partial\left(u_i u_j\right)}{\partial x_j}=F_i-\frac{1}{\rho}\frac{\partial p}{\partial x_i}+\nu\frac{\partial^2 u_i}{\partial x_j^2}+e_z \quad i=1,2,\cdots,\mathrm{n} \tag{9.6}$$

where the index i expresses the component and the index j expresses the addition index (j = 1, 2, and 3).

Flow in the hydrodynamic machine is 3D and analytical solutions of the Navier–Stokes equations describing flow in a hydrodynamic machine with complicated 3D geometry have not yet been found. Therefore, numerical methods are used. In CFD simulation, the internal space of the hydrodynamic machine must be divided into a finite number of volumes (if the finite-volume method is applied), to which the Navier–Stokes (and other) equations are subsequently applied.

9.3 TURBULENCE MODELING METHODS

The flow in the hydrodynamic machine is characterized by high values of the Reynolds number and is therefore highly turbulent. The instantaneous values of the quantities in the flow field change with very high frequency. The structure of the flow field is also complicated as it includes boundary layer detachment. The aim of CFD simulation is to realize precise calculation of quantities in the flow field, including flow separation and eddies. One of the possibilities is the application of direct numerical simulation (DNS). The DNS of turbulence involves solving the Navier–Stokes equations numerically with physically consistent precision in space and time, resolving all of the important turbulent scales. DNS tries to solve the Navier–Stokes equations, which control fluid flows, directly without the need for any modeling assumptions. A precise 3D time-dependent solution of the governing equations may be obtained if the mesh is fine enough, the time step is short enough, and the numerical technique is tailored to minimize dispersion and dissipation errors. The only mistakes are those produced by residual approximations contained in the numerical scheme and the computing machine's number representation technology. A solution of the Navier–Stokes equations as a numerical mathematics exercise differs fundamentally from a solution of the same equations with the goal of obtaining a precise correlation of the results with turbulence physics, in that the accuracy of the calculations must be closely monitored in the latter case. The significant amount of processing resources necessary to execute turbulence computations at Reynolds numbers approaching that of practical interest is the main issue. As a result of this reality, only basic flows have been studied with DNS for a long time.

The second option for CFD simulation of turbulent flow is an approach based on direct calculation of large eddy simulation (LES). The smallest turbulence scales are spatially filtered out in LES, whereas the biggest, most energy-containing scales are directly resolved. Small-scale vortices are modeled with the application of a suitable turbulence model. This approach is also demanding on the density of the computational mesh and thus also on the performance of the computer.

A third option for calculating turbulent flow in a hydrodynamic machine is to apply Reynolds averaged Navier–Stokes (RANS) equations. The RANS equations apply the decomposition of instantaneous values of flow field quantities, which change with extremely high frequency to time-averaged values (denoted by a line above the relevant quantity) and fluctuation components (denoted by an apostrophe), as shown in Equation (9.7). The presence of fluctuating components in the RANS equations causes additional shear stresses called Reynolds turbulent stresses. The resulting RANS form for a stationary coordinate system can be expressed according to Equation (9.7) (body forces are neglected).

$$\frac{\partial \overline{u'_i u'_j}}{\partial t} + \overline{u}_k \frac{\partial \overline{u'_i u'_j}}{\partial x_k} = -\underbrace{\frac{\partial}{\partial x_k}\left[\left(\overline{u'_i u'_j u_k'}\right) + \overline{\frac{p'}{\rho}\left(\delta_{kj} u'_i + \delta_{ik} u'_j\right)} - \nu \frac{\partial}{\partial x_k}\left(\overline{u'_i u'_j}\right)\right]}_{diffusion} -$$

$$-\underbrace{\left[\overline{u'_i u'_k}\frac{\partial \overline{u}_j}{\partial x_k} + \overline{u'_j u'_k}\frac{\partial \overline{u}_i}{\partial x_k}\right]}_{production} + \underbrace{\overline{\frac{p'}{\rho}\left[\frac{\partial u'_i}{\partial x_j} + \frac{\partial u'_j}{\partial x_i}\right]}}_{redistribution} - \underbrace{2\nu \overline{\left[\frac{\partial u'_i}{\partial x_k}\frac{\partial u'_j}{\partial x_k}\right]}}_{dissipation} \tag{9.7}$$

9.4 REYNOLDS EQUATIONS

Reynolds turbulent stresses form a tensor of nine terms, with six independent members. This system is an extensive set of differential equations that are difficult to solve. For this reason, theories have been developed that deal with simplification or expression of Reynolds stresses. RANS-based turbulence models are based on these theories. The task of each turbulence model is then to express the turbulent stress, or other equivalent scalar quantities, and then determine the distribution of this parameter in the flow field.

9.4.1 RANS Turbulence Models

RANS turbulence models are used to solve averaged Navier–Stokes equations. These models are based on different approaches to solving Reynolds stresses. Some of the RANS turbulence models are shown in Figure 9.1.

The first and most widespread subset of RANS turbulence models includes models based on an approach that applies the Boussinesq hypothesis to express Reynolds stresses. This hypothesis assumes that turbulent stresses and flows are proportional to the mean velocity and temperature gradient. This subset of models was named "Eddy viscosity models." The Reynolds stress tensor in the time-averaged Navier–Stokes equation is replaced by the turbulent viscosity multiplied by the velocity gradients in eddy viscosity turbulence models. This relationship is known as the Boussinesq assumption, where the Reynolds stress tensor in the time-averaged Navier–Stokes equation is replaced by the turbulent viscosity multiplied by the velocity gradients.

9.4.1.1 Eddy Viscosity Models

"Eddy viscosity" turbulence models are focused on the solution of turbulent viscosity. The turbulent viscosity (μ_t dynamic turbulent viscosity) represents the proportionality constant on which the magnitude of the additional turbulent stresses arising from

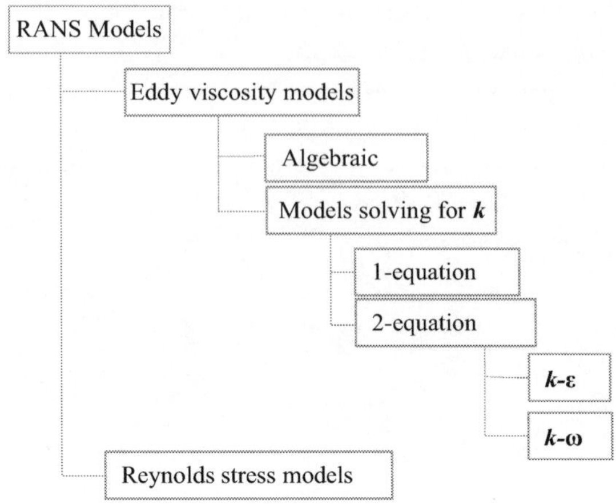

FIGURE 9.1 Classification of Reynolds averaged Navier–Stokes equations (RANS) models.

the fluctuating velocity components depends. Turbulent viscosity is not a property of the flowing fluid, but a characteristic of the flow and depends on the location in the flow field. In general, for the additional shear stress (due to fluctuating velocity components) in the *xy* plane, Equation (9.8) applies. The velocity component in the *x* direction is denoted by *u*, and the component in the *y* direction is denoted by *v*.

$$\tau_t = -\rho\overline{u\text{v}} = \mu_t\left(\frac{\partial \overline{u}}{\partial y}\right) \tag{9.8}$$

Next, we can write a relation for the expression of the fluctuation components of velocity in the form of Equation (9.9).

$$-\overline{u'_i u'_j} = \nu_t\left(\frac{\partial \overline{u_i}}{\partial x_j} - \frac{\partial \overline{u_j}}{\partial x_i}\right) - \frac{2}{3}k\delta_{ij} \tag{9.9}$$

where

$k = \frac{1}{2}\overline{u'_i u'_i}$ is turbulent kinetic energy (TKE)
ν_t is turbulent kinematic viscosity
$\mu_t = \rho\nu_t$.

Note that Equation (9.9) takes into account the compressibility of the fluid.

According to the number of equations on which the calculation of turbulent viscosity is performed, this subgroup of models is divided into other subgroups. These are zero-equation (or algebraic), one-equation, and two-equation models. One-equation and two-equation models are together classified as a group of models applying TKE in the calculation of turbulent viscosity. These turbulence models are characterized by a description of the local state of turbulence using a velocity and length scale, which are directly related to the turbulent viscosity.

The second important subgroup of turbulence models includes the models referred to as "Reynolds stress models." The Boussinesq assumption is not utilized in Reynolds stress models; instead a partial differential equation (transport equation) for the stress tensor is obtained from the Navier–Stokes equation.

9.4.1.2 Algebraic (Zero-Equation) Models

Algebraic models apply simple algebraic relations in the calculation of turbulent viscosity. The first of these models is the Prandtl model from 1925. The model was based on observations of turbulent flow, in which visually detected "clusters" of particles were carried by the flow. The clusters remained together for a certain time during the flow and later they were dispersed and became extinct. The distance that these particles traveled without contact with other clusters is called the "mixing length," l_{mix}. The speed at which they move is called the mixing speed, v_{mix}. Analogous to the kinetic theory of gases, Prandtl introduced Equation (9.10), replacing the gas molecule with the observed particle cluster and replacing the mean free path of the molecule with the mixing length. For simplicity, a relation for 2D flow is illustrated in Figure 9.2.

$$\tau_t = \frac{1}{2}\rho v_{mix} l_{mix} \frac{\partial \bar{u}}{\partial y} = \mu_t \frac{\partial \bar{u}}{\partial y} \tag{9.10}$$

The unknown mixing speed is given by Equation (9.11).

$$v_{mix} = const.l_{mix} \left| \frac{\partial \bar{u}}{\partial y} \right| \tag{9.11}$$

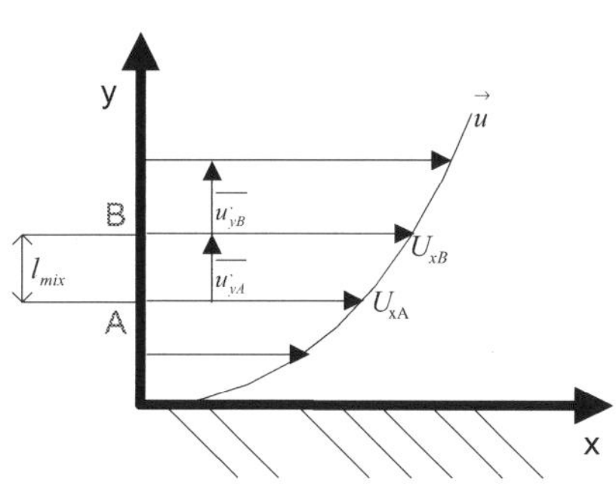

FIGURE 9.2 Velocity profile and Prandtl's mixing length theory.

After substituting Equation (9.11) into Equation (9.10), we obtain the resulting relation, which is used in the model to calculate the dynamic turbulent viscosity as given in Equation (9.12).

$$\mu_t = \rho l_{mix}^2 \left| \frac{\partial \bar{u}}{\partial y} \right| \tag{9.12}$$

The kinematic turbulent viscosity is given by:

$$\nu_t = l_{mix}^2 \left| \frac{\partial \bar{u}}{\partial y} \right| \tag{9.13}$$

In the case of shear layer flow modeling, it is possible to determine the mixing length according to Equation (9.14). κin Equation (9.14) is Von Kármán's constant and y represents the perpendicular distance from the wall.

$$l_{mix} = \kappa y \tag{9.14}$$

The model can also be applied to model a developed flow under certain circumstances. In 2D flow, it can be applied to a flow with a small flow separation without large pressure gradients and a small flow curvature. In 3D flow, it can be used if the flow is characterized by small secondary flow and small flow separation.

9.4.1.3 Models Based on Turbulent Kinetic Energy

In zero-equation models, from dimensional analysis, the turbulent kinematic viscosity is given as the product of the characteristic length (l_k) and velocity scale (u_k).

$$\nu_t = \frac{\mu_t}{\rho} = l_k u_k \tag{9.15}$$

where:

l_k is the characteristic length of the largest vortices
u_k is the characteristic velocity of the largest vortices.

Models using TKE are based on the assumption that the characteristic velocity scale, u_k, is proportional to TKE.

$$u_k = const \sqrt{k} \tag{9.16}$$

where $k$is the TKE of the fluctuating velocity components per unit mass. For TKE (k), Equation (9.17) generally applies.

$$k = \frac{1}{2}\left(\overline{u'^2_1} + \overline{u'^2_2} + \overline{u'^2_3}\right) = \frac{1}{2}\overline{u'^2_i} \tag{9.17}$$

Equation (9.17) is based on dimensional analysis of the velocity scale and kinetic energy (k) is proportional to velocity, i.e., $[u_k]^2$. The turbulent viscosity is then determined according to the Kolmogorov–Prandtl relation as follows:

$$\nu_t = \frac{\mu_t}{\rho} = C_\nu \sqrt{k}l \tag{9.18}$$

where C_ν is empirical constant.

It is possible to derive a transport equation for TKE. Depending on whether the flow length scale is determined on the basis of empirical relationships (similarly to the case of zero-equation models) or on the basis of another transport equation, we speak of one-equation or two-equation models. The transport equation for TKE has the form of Equation (9.19).

$$\begin{aligned} \underbrace{\frac{\partial k}{\partial t}}_{I} + \underbrace{\frac{\partial \overline{u_j k}}{\partial x_j}}_{II} &= \underbrace{-\frac{\partial}{\partial x_j}\left[\overline{u'_j\left(\frac{u'_i u'_i}{2} + \delta_{ji}\frac{p'}{\rho}\right)}\right]}_{III} + \\ &\underbrace{\nu\frac{\partial^2 k}{\partial x_j^2}}_{IV} \underbrace{- \overline{u'_i u'_j}\frac{\partial \overline{u'_i}}{\partial x_j}}_{V} \underbrace{- \nu\overline{\frac{\partial u'_i}{\partial x_j}\frac{\partial u'_i}{\partial x_j}}}_{VI} \end{aligned} \tag{9.19}$$

where:

- I is change in velocity
- II is convection
- III is turbulent diffusion due to velocity and pressure fluctuations
- IV is molecular diffusion
- V is production: large turbulent scales extract energy from the mean flow
- VI is viscous dissipation: transformation of kinetic energy at small scales to internal energy.

After further modifications and substitutions, we get to Equation (9.20).

$$\frac{\partial k}{\partial t} + \frac{\partial \bar{u}_j k}{\partial x_j} = \frac{\partial}{\partial x_j}\left(\frac{\nu_t}{\sigma_k}\frac{\partial k}{\partial x_j}\right) + \underbrace{\nu_t\left(\frac{\partial \bar{u}_j}{\partial x_i} + \frac{\partial \bar{u}_i}{\partial x_j}\right)\frac{\partial \bar{u}_i}{\partial x_j}}_{P} \underbrace{- C_D\frac{k^{\frac{3}{2}}}{l}}_{\varepsilon} \tag{9.20}$$

where $\varepsilon$is the dissipation of specific kinetic turbulent energy and σ_k and C_D are empirical constants. The last term in Equation (9.20) (ε) can be determined as follows.

$$\varepsilon = \nu \overline{\frac{\partial u'_i \partial u'_i}{\partial x_j^2}} = C_D \frac{k^{\frac{3}{2}}}{l} \tag{9.21}$$

9.4.1.4 One-Equation Models

One-equation models apply the Kolmogorov–Prandtl Equation (9.18) to determine kinematic turbulent viscosity. Equation (9.18) is used to calculate the turbulent kinematic viscosity, but the calculation of the length scale is based on empirical relationships, like the zero-equation models.

9.4.1.5 Two-Equation Models

In two-equation models, unlike 1D models, another transport equation is used to calculate a quantity that is equivalent to the length scale.

9.4.1.5.1 k-ε Model

The standard *k*-ε model applies proportionality, Equation (9.22) to determine turbulent viscosity and Equation (9.23) to the length scale of turbulence.

$$\mu_t \approx \frac{\rho k^2}{\varepsilon} \tag{9.22}$$

$$l \approx \frac{k^{\frac{3}{2}}}{\varepsilon} \tag{9.23}$$

In this model, turbulent viscosity is determined using two transport equations for *k* and ε. To calculate dynamic turbulent viscosity, Equation (9.24) is used.

$$\mu_t = \rho C_\mu \frac{k^2}{\varepsilon} \tag{9.24}$$

$$\rho \frac{\partial k}{\partial t} + \rho \bar{u}_j \frac{\partial k}{\partial x_j} = \mu_t \left(\frac{\partial \bar{u}_j}{\partial x_i} + \frac{\partial \bar{u}_i}{\partial x_j} \right) \frac{\partial \bar{u}_i}{\partial x_j} - \rho \varepsilon + \frac{\partial}{\partial x_j} \left(\left(\mu + \frac{\mu_t}{\sigma_k} \right) \frac{\partial k}{\partial x_j} \right) \tag{9.25}$$

To determine TKE, either Equation (9.20) or Equation (9.25) can be used. The dissipation of a specific kinematic turbulent energy,ε, can be expressed from the Navier–Stokes equations as follows.

$$\rho\frac{\partial\varepsilon}{\partial t}+\rho\bar{u}_j\frac{\partial\varepsilon}{\partial x_j}=C_{\varepsilon 1}\frac{\varepsilon}{k}\mu_t\left(\frac{\partial\bar{u}_j}{\partial x_i}+\frac{\partial\bar{u}_i}{\partial x_j}\right)\frac{\partial\bar{u}_i}{\partial x_j}-$$

$$-C_{\varepsilon 2}\frac{\varepsilon^2}{k}+\frac{\partial}{\partial x_j}\left(\left(\mu+\frac{\mu_t}{\sigma_\varepsilon}\right)\frac{\partial\varepsilon}{\partial x_j}\right) \tag{9.26}$$

$$\omega=\frac{\varepsilon}{C_\mu k},\ l=\frac{C_\mu k^{\frac{3}{2}}}{\varepsilon} \tag{9.27}$$

The empirical coefficients in Equation (9.24)–Equation (9.27) take the values expressed in Equation (9.28).

$$C_{\varepsilon 1}=1.44,\cdots,C_{\varepsilon 2}=1.92\cdots,C_\mu=0.09,\cdots,\sigma_k=1 \tag{9.28}$$

9.4.1.5.2 k-ω Model

A modern version of the k-ω model for calculating turbulent quantities is derived into the following form.

$$\rho\frac{\partial k}{\partial t}+\rho u_j\frac{\partial k}{\partial x_j}=\tau_{ij}\frac{\partial u_i}{\partial x_j}-\beta^*\rho k\omega+\frac{\partial}{\partial x_j}\left(\left(\mu+\sigma^*\mu_t\right)\frac{\partial k}{\partial x_j}\right) \tag{9.29}$$

$$\rho\frac{\partial\omega}{\partial t}+\rho u_j\frac{\partial\omega}{\partial x_j}=\alpha\frac{\omega}{k}\tau_{ij}\frac{\partial u_i}{\partial x_j}-\beta\rho\omega^2+\frac{\partial}{\partial x_j}\left(\left(\mu+\sigma\mu_t\right)\frac{\partial\omega}{\partial x_j}\right) \tag{9.30}$$

Empirical constants and complementary relations used are given below:

$$\alpha=\frac{5}{9},\ \beta=\frac{3}{40},\quad \sigma=0.5,\ \sigma^*=0.5 \tag{9.31}$$

$$\varepsilon=\beta^*\omega k,\ l=\frac{\sqrt{k}}{\omega} \tag{9.32}$$

Dimensional analysis results in relationships that can be used to express the equation for ω:

$$\mu_t\approx\frac{\rho k}{\omega} \tag{9.33}$$

$$l \approx \frac{\sqrt{k}}{\omega} \tag{9.34}$$

$$\varepsilon \approx \omega k \tag{9.35}$$

9.4.1.5.3 Shear Stress Transport (SST) Model

The SST model was developed by Menter in 1994. The motivation for its creation was the need to simulate flow characterized by a strong inverse-pressure gradient and flow separation (boundary layer separation) from the walls. This model takes the advantages or compensates for the shortcomings of the *k*-ε and *k*-ω models. The standard *k*-ε model is universal, stable, and robust, but it insufficiently captures phenomena such as boundary layer separation from the wall and fluid flow in the boundary layer. The *k*-ω model better captures the flow behavior in the boundary layer and provides relatively accurate flow results with a medium inverse pressure gradient. However, the *k*-ω model is too sensitive to values of ω in the free-flow region outside the boundary layer. In this flow region, it is more advantageous to model turbulence using the *k*-ε model.

Therefore, the SST model combines turbulence models *k*-ε and *k*-ω and applies the suitable turbulence model to a given flow region (flow zone). The zonal formulation of the SST model is based on the so-called blending functions that ensure the correct selection of *k*-ε and *k*-ω zones without user intervention. When using the SST model, it is necessary to calculate the distance from the wall, *y*, which is essential for the calculation of the connection function. The calculation of the distance from the wall is based on the so-called Poisson's equation.

When using the SST model, it is possible to apply the *k*-ε model from Equations (9.25) and (9.26) in the determination of TKE and TKE dissipation rate (ε).

The *k*-ω model uses a specific value of ω instead of the rate of dissipation of TKE. In the *k*-ω model, for turbulent viscosity and the length scale of turbulence, the proportionality Equations (9.33) and (9.34) apply.

The SST model is based on the equations of the *k*-ω model; however the calculation of the turbulent viscosity is determined using Equation (9.36).

$$\nu_t = \frac{a_1 k}{\max\left(a_1 \omega, S, F_2\right)} \tag{9.36}$$

$$\mu_t = \rho \nu_t \tag{9.37}$$

The quantity S in Equation (9.36) is the invariant measure of shear stress and F_2 is the blending function given by Equation (9.42) (see below). The equation for TKE takes the form of Equation (9.38) and the equation for the specific dissipation of TKE has the form of Equation (9.39).

$$\rho\frac{\partial k}{\partial t}+\rho\frac{\partial (u_i k)}{\partial x_i}=\tilde{P}_k-\beta^*\rho k\omega+\frac{\partial}{\partial x_i}\left[(\mu+\sigma_k\mu_t)\frac{\partial k}{\partial x_i}\right] \tag{9.38}$$

$$\rho\frac{\partial \omega}{\partial t}+\rho\frac{\partial (u_i \omega)}{\partial x_i}=\alpha\rho S^2-\beta\rho\omega^2+\frac{\partial}{\partial x_i}\left[(\mu+\sigma_\omega\mu_t)\frac{\partial \omega}{\partial x_j}\right]+ \\ +2(1-F_1)\rho\sigma_{\omega 2}\frac{1}{\omega}\frac{\partial k}{\partial x_i}\frac{\partial \omega}{\partial x_i} \tag{9.39}$$

where
α, β, σ_ω, $\sigma_{\omega 2}$ are constants. Parameter F_1 is a blending function and given by Equation (9.40).

$$F_1=tgh\left\{\left\{min\left[max\left(\frac{\sqrt{k}}{\beta^*\omega y},\frac{500\nu}{y^2\omega}\right)\frac{4\rho\sigma_{\omega 2}k}{CD_{k\omega}y^2}\right]\right]^4\right\} \tag{9.40}$$

Equation (9.41) is used to determine $CD_{k\omega}$, and y represents the distance to the nearest wall.

$$CD_{k\omega}=max\left(2\rho\sigma_{\omega 2}\frac{1}{\omega}\frac{\partial k}{\partial x_i}\frac{\partial \omega}{\partial x_i}10^{-10}\right) \tag{9.41}$$

$$F_2=tgh\left[\left[max\left(\frac{2\sqrt{k}}{\beta^*\omega y},\frac{500\nu}{y^2\omega}\right)\right]^2\right] \tag{9.42}$$

$$P_k=min\left(10P_k\beta^*\rho k\omega\right) \tag{9.43}$$

$$P_k=\mu_t\frac{\partial U_i}{\partial x_j}\left(\frac{\partial U_i}{\partial x_j}+\frac{\partial U_j}{\partial x_i}\right) \tag{9.44}$$

The constants α and β are calculated by a blend from the corresponding constants of k-ε and k- ω using Equation (9.45).

$$\alpha=\alpha_1 F+\alpha_2(1-F) \tag{9.45}$$

The values of these constants are given in Table 9.1.

TABLE 9.1
Constants of the Shear Stress Transport (SST) Turbulence Model

α1	α2	β1	β2	β*	σk1	σk2	σω2	σω1
5.9	0.44	3.40	0.0828	9/100	0.85	1	0.856	1.2

It should be noted that currently the SST model is often used in standard CFD simulations. Alternatively, the scale adaptive simulation SST model is often applied.

In addition to the turbulence models mentioned, there are a number of other models that are currently commonly implemented in the available CFD codes.

9.4.1.6 Modeling Flow Near Wall

The region of the boundary layer can be divided as follows. In the immediate vicinity of the wall there is a laminar sublayer, the outer part of the boundary layer is called a fully developed turbulent boundary layer, and between them there is a transition sublayer (Figure 9.3).

In the region of the laminar sublayer, the flow is laminar, and the molecular viscosity has a dominant effect. In the area of the turbulent sublayer, the influence of turbulent stresses is dominant, and the influence of dynamic viscosity is smaller. There is a significant production of TKE due to high Reynolds stresses and a medium-velocity gradient. In the transition sublayer, the effect of dynamic viscosity as well as turbulent stresses is at a similar level.

The topic of wall functions was analyzed, for example, in the literature. In general, it should be noted that the issue of flow modeling near the wall is a complex topic. Several approaches to modeling the flow near the wall region were developed. It is beyond the scope of this book to take a detailed look at the individual approaches here. Commercial CFD codes have successfully implemented a wide range of near-wall flow models. For these reasons, we limit ourselves to a brief description of the main approaches to flow modeling near the wall.

Flow analysis near the wall can be performed in two ways. The first way is to apply wall functions, which represent a set of semi-empirical relationships. The wall functions are applied to determine quantities in the flow field in the laminar sublayer and the transition sublayer. The second way is to model the flow near the wall using a very fine computational mesh. The computational mesh must be fine enough to ensure the correct calculation of velocity gradients, shear stresses, as well as other quantities in the flow fields in the laminar and transition sublayers.

9.4.1.6.1 Wall Functions

Standard wall functions can be derived from the momentum equation. From the Navier–Stokes equation, after simplifications, we get Equation (9.46). In Equation (9.46), the following simplifications are considered:

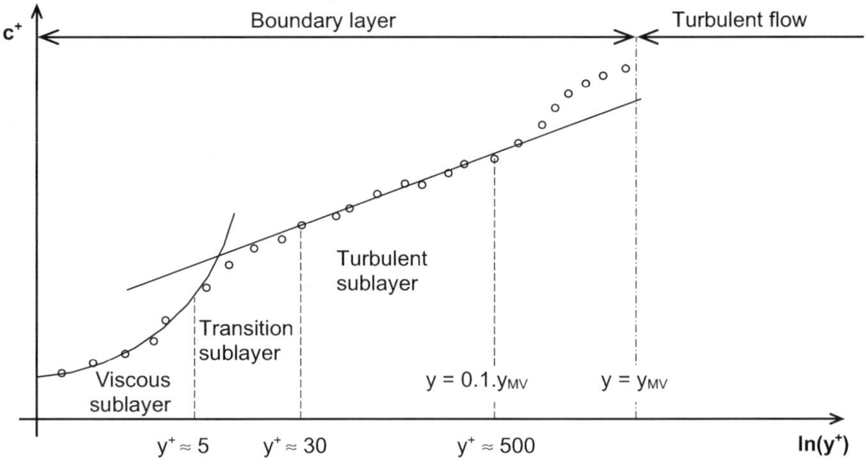

FIGURE 9.3 Boundary layer near the wall.

- Flow is fully developed due to pressure gradients, with zero velocity and stress gradients in the main direction of the flow (*x*-axis – parallel to the wall).
- The velocity in the direction normal to the wall (*y*-axis) is zero.
- The change of any quantity in the *z*-axis direction is zero.

$$\frac{\partial}{\partial y}\left(\mu\frac{d\bar{c}}{dy}-\rho\overline{u'v'}\right)=0 \tag{9.46}$$

After applying boundary conditions at the wall, Equation (9.47) and integrating Equation (9.46), we get the relation for the shear stress on the wall in form of Equation (9.48).

$$\text{For }\ y=0,\quad \overline{u'\text{v}'}=0 \tag{9.47}$$

$$\tau_w=\mu\frac{d\bar{c}}{dy}-\rho\overline{u'\text{v}'} \tag{9.48}$$

Note that the parameter y^+ expresses the dimensionless distance from the wall and is given by Equation (9.49). In the laminar sublayer ($0 < y^+ < 5$), where no turbulent stresses occur, Equation (9.50) applies to the shear stress on the wall in the direction of the flow.

$$y^+=\frac{\rho u_\tau y}{\mu} \tag{9.49}$$

where

u_τ is friction velocity, $u_\tau = \sqrt{\frac{\tau_w}{\rho}}$. u_τ is termed the friction velocity because it has dimension m/s.

y is the distance of point P from the wall

$$\tau_w = \mu \frac{\partial u}{\partial y} \tag{9.50}$$

After integrating Equation (9.48), we obtain Equation (9.51) for velocity as a function of the perpendicular distance from the wall (y).

$$\frac{u}{u_\tau} = \frac{y u_\tau}{\nu} \tag{9.51}$$

The mean flow velocity in the laminar sublayer is given by Equation (9.52).

$$u^+ = \frac{u}{u_\tau} \tag{9.52}$$

At the same time, for a laminar sublayer, the following relation applies, i.e., the mean velocity magnitude is equal to the value of y^+.

$$u^+ = y^+ \tag{9.53}$$

Note that as the mean velocity rises, the viscous sublayer becomes thinner. As a result, at very high Reynolds numbers, the velocity profile becomes almost flat, and the velocity distribution becomes more uniform (very low viscosity).

At a distance of $30 < y^+ < (y / y_{MV} \approx 0.1)^+$ from the wall, the influence of viscous stresses is small and a turbulent stress begins to prevail. Equation (9.48) can therefore be written in the form of Equation (9.54).

$$\rho \overline{u'v'} = \tau_w \tag{9.54}$$

Mathematically, the relation for the flow near the wall can be derived using the mixing length hypothesis, Equation (9.10) and Equation (9.12). For the stress we then get the relation Equation (9.55).

$$\tau_w = \tau_{xy} = \rho l_{mix}^2 \left(\frac{\partial U}{\partial y} \right)^2 = \rho \kappa y^2 \left(\frac{\partial U}{\partial y} \right)^2 \tag{9.55}$$

where κ is von Karman's constant and y is the distance from the wall. We can get the relationship for velocity near the wall from Equation (9.55) in the form of Equation

(9.56), which is known as the "law of the wall." It describes the relationship between speed and stress on the wall.

$$U = \frac{1}{\kappa}\sqrt{\frac{\tau_w}{\rho}} \ln(y) + const \tag{9.56}$$

Equation (9.56) can be rewritten as follows. The value of B is given in Table 9.2 based on the experimental Nikuradze, where the y_0 term expresses the distance from the wall at which the ideal velocity given by Equation (9.57) goes to zero.

$$U = \sqrt{\frac{\tau_w}{\rho}}\left(\frac{1}{\kappa}\ln\left(\frac{y}{y_0}\right) + B\right) \tag{9.57}$$

Using Equation (9.57) it is possible to derive the wall functions for individual transport quantities in turbulence models. These are given in Equation (9.58).

$$k = \frac{\sqrt{\frac{\tau_w}{\rho}}}{\sqrt{\beta^*}} \qquad \omega = \frac{\sqrt{k}}{\sqrt[4]{\beta^*}\kappa y} \qquad \varepsilon = \left(\beta^*\right)^{3/4}\frac{k^{3/2}}{\kappa y} \tag{9.58}$$

The condition for the application of wall functions is that the first node of the computational mesh is located in the area of predominance of turbulent stresses, where the values of the dimensionless distance from the wall are in the range $30 < y^+ < 60$.

9.4.1.6.2 *Direct Modeling of Flow in Low Reynolds Number Region*

In this case, the flow is modeled in the laminar, transition, and turbulent sublayer. In order to be able to apply flow modeling near the wall, at least ten layers of computational mesh cells must lie in the area of the laminar sublayer, and the value of the dimensionless distance from the wall (y^+) should satisfy the condition $y^+ < 2$.

Some current CFD codes apply the so-called "automatic wall function" in conjunction with the k-ω-based turbulence model. In this approach, suitable equations for calculation of the flow field quantities are automatically selected depending on the fineness of the computational mesh.

TABLE 9.2
Values of Constant *B* for Different Turbulence Models

Model	*B*
***k*-omega (Kolmogorov)**	3.1
***k*-omega (Wilcox)**	5.1
***k*-epsilon**	–2.2
Experiment (Nikuradze)	5

At the end of this section we would like to note that our goal was not to describe the full breadth of mathematical modeling of 3D turbulent flow. Our goal was only to summarize simulation approaches and mathematical description of the flow of a real liquid using CFD. The reader is advised to seek further reading such as books by Wilcox for detailed information on this subject.

At this point, we also emphasize that it is practically impossible to apply the above models in the direct design of a pump and turbine geometry. Simplifications of the flow model are usually used for practical application in the direct design of hydrodynamic machine geometry. Of course, the accepted simplifications introduce smaller or larger inaccuracies into the solution. Ultimately the proposed primary geometry must be subjected to further modifications. Before that, however, we perform a simulation on the primary geometry of the CFD using the above-mentioned flow models, in order to more accurately determine the performance parameters. Subsequently, we perform the necessary modifications to the primary geometry.

9.4.2 Rotor–Stator Interface

The computational domain usually consists of blocks that represent either the rotating or nonrotating part of the hydrodynamic machine. The computational meshes of these blocks are generated separately and there is usually no conformal interface between them. This means that the position of the individual nodes of the computational mesh at the interface of two mutually adjacent blocks is not identical. Therefore, it is necessary to ensure the transmission (flow) of the flow field quantities between the cells of the computational meshes of adjacent blocks of the computational domain.

The transfer of flow field quantities between two blocks of a computational mesh that have a nonconforming mesh interface (they can be either nonrotating or one rotating and the other nonrotating) can be realized on the basis of the following model. A virtual surface with zero thickness is located between the boundary zones of the blocks of the computational mesh, as shown in Figure 9.4. Within this virtual area (called the control areas) it is assumed that the flow of quantities is constant.

The control area must be divided into a set of smaller control areas (CS). Each node of the computational meshes of both blocks of the computational area must have a corresponding node on the control surface, which has the same position as shown in Figure 9.5. The nodes on the control area then demarcate the smaller control areas. Each cell of the computational mesh of both neighboring domains has its own boundary integration point (IP) located at the edge of the cell, as shown in Figure 9.4, which touches the control area. The quantities related to this point are transmitted through the control area to the nearest integration points of the computational mesh of the second domain.

In this transmission of flow field, it is first necessary to calculate the sizes of the individual areas of the control area (A_{SC}), which are the intersection of two partially contacting cell walls of the computational meshes of both domains ("Area ip" or "A_{IP}" in Figure 9.5). Then, the ratios of the sizes of the control areas and the sizes of the adjacent cell wall contents are calculated ($F_{IP,CS}$ in Figure 9.5). For flow of some quantity $\emptyset$ from cell 1, which is represented by the integration point IP1, Equation (9.59)

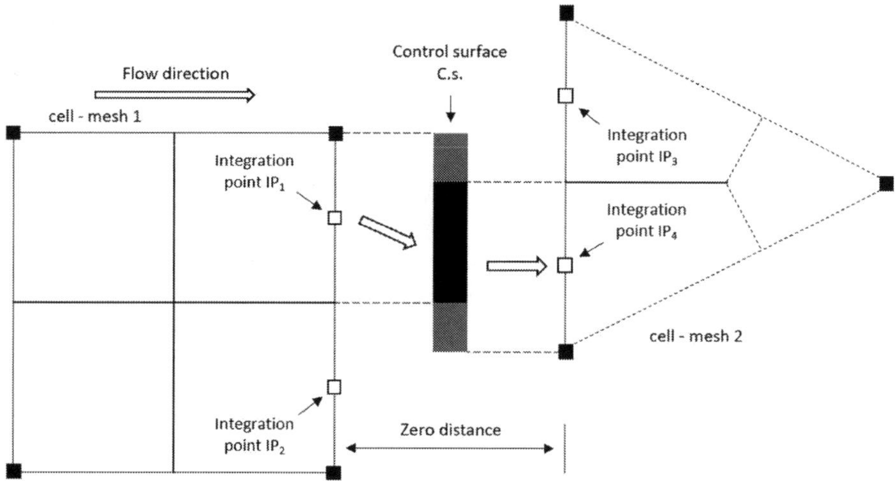

FIGURE 9.4 Scheme of transfer of flow field quantities through the interface.

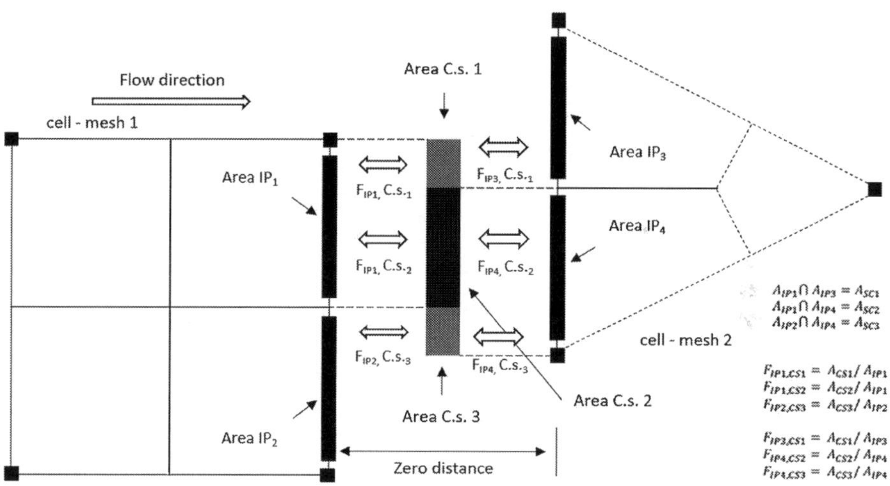

FIGURE 9.5 Schematic of determination of interface area ratios.

applies. Equation (9.60) applies to the flow of this quantity through the control surface CS2. $F_{\mathrm{IP,\,CS2}}$ is the ratio of the sizes of the control area CS2 and the area of the cell wall to which the integration point IP1 applies (Figure 9.4). Thus, the resulting flow of the quantity $\varnothing$ from the integration point IP1 to the control area CS2 will be recalculated based on the ratio of the size of the control area CS2 and the wall size of the cell to which the integration point IP1 applies. For the flow of the quantity through the control surface CS2, Equation (9.61) also applies.

$$Flux_{IP1} = \dot{m}_{IP1}\Phi_{IP1} \tag{9.59}$$

$$Flux_{CS2} = F_{IP1,CS2}Flux_{IP1} \tag{9.60}$$

$$Flux_{CS2} = F_{IP4,CS2}Flux_{IP4} \tag{9.61}$$

A similar principle can be applied to the case of two adjacent blocks, which are both stationary or both rotating. With this type of interface, the transfer of flow field quantities between adjacent blocks is realized by the above-mentioned principles without modifications.

In general, the above procedure can be implemented in several ways. The first way is the "frozen rotor" interface applicable to time-steady flow. The flow field variables are handled in this type of interface in such a way that the transition from the rotating to the stationary part changes the coordinate system (from rotating to stationary) while maintaining the original relative position of the rotor relative to the stator.

The periodicity of the geometry of the impeller and the diffuser can be used to reduce the number of channels included in the calculation, provided that the pitch angle of the rotor and stator domains is approximately the same. Then the fluid flow at the interface is also calculated according to a scale corresponding to the ratio of the pitch angles of the stator and rotor. The disadvantage of this approach is that the result depends on the relative position of the rotor and the stator. The advantage is its robustness and relatively less computational time.

The second way is a "mixing plane" interface. In realistic turbomachinery simulations, a mixing plane interface offers a circumferentially averaging rotor–stator coupling interface, which is extremely helpful. The individual quantities of the flow field are averaged around the circumference of the outlet zone of one of the components of the computational mesh and these average values are transferred to the inlet zone of the next component downstream. This approach can be applied in both steady and unsteady flows. In the case of unsteady flow, it is possible to use the "transient totor–stator" interface. For uneven blade counts and pitches of neighboring blade rows, the transient rotor–stator technique scales the circumferential flow distribution. As a result, the flow distribution is stretched or compressed at the interfaces in the circumferential direction and in relation to the blade pitches. This type of scaling causes a transformation error at the interfaces, resulting in blade-passing frequency variations. In this approach, for each time step, the instantaneous position of the rotor relative to the stator is considered; this corresponds to the rotation of the rotor in one time step. The model takes into account in detail the interaction of the impeller and the stator and thus the calculated flow field is more similar to real, but its use is time consuming and also demanding on hardware.

9.5 CFD SIMULATION WORKFLOW

CFD simulation consists of preparation of geometry (space in which fluid flows), generation of computational mesh, specification of CFD simulation methods (selection of solver, turbulence model), specification of boundary and initial conditions, simulation itself during which it is necessary to monitor convergence and stability of solution, and evaluation of results (postprocessing).

9.5.1 CFD Computational Domain

The boundaries of the computational domain must be defined with respect to the required result of the CFD simulation. If the effort is only to determine the main performance parameters of the impeller, it is sufficient to define a computational domain with boundaries near the inlet and outlet of the impeller, while the simulation is sufficient only in one channel using the so-called periodic boundary conditions. If it is necessary to analyze in detail the performance parameters of the machine (which are often significantly affected by the mutual interactions between the rotor and stator), it is necessary to define the boundaries of the computational domain at a sufficient distance in front of the inlet body or behind the outlet. For a more detailed simulation that considers the interaction of the rotor and the stator, the simulation is carried out in all channels (without applying periodic boundary conditions). In the case of impellers without a front shroud, the flow in the gaps between the tips of the blades and the stator body should also be considered.

In the inflow and outflow domains, the general assumption is that there is a perpendicular orientation (to the flow area) of the velocity vectors. At the inlet the vectors point inwards and at the outlet of the computational domain they point outwards. It is known that, especially before entering the impeller, intense secondary flows can occur in high-speed machines, which extend a considerable distance in front of the impeller. These secondary flows have a significant effect on the internal structure of the flow field in the impeller, which is also reflected in the performance parameters. If the inlet section is too close to the impeller, the simulation of the secondary flows in front of the impeller could be disturbed by the fixed orientation of the velocity vectors in the inlet because of boundary conditions. A similar situation applies to the region behind the outlet. Therefore, the computational domain should be properly defined. Examples of proper definitions of the computational domain of hydrodynamic machines are shown in Figures 9.6 to 9.9.

9.5.2 Computational Mesh

The generation of the computational mesh is realized in the space around one blade of the given machine. This creates one computational mesh block representing one passage between blades. The composition of several such blocks will create a simulation mesh of the entire domain of the component of the machine. Performance of machines, head losses for internal flows, and drag force for exterior flows are generally predicted more accurately using structured hexahedral cells. Hence computational meshes consisting of hexahedral cells are preferred. Creating a high-quality hexahedral mesh for complicated geometry takes a significant amount of work, ranging from a few days to a few weeks. Furthermore, for highly complicated geometries, obtaining a mesh of adequate quality may be difficult. In some cases, such as

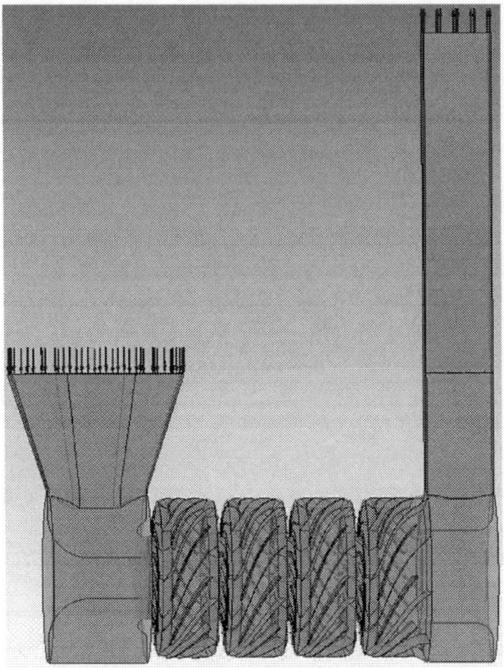

FIGURE 9.6 Computational mesh of a multi-stage pump.

FIGURE 9.7 Computational mesh of a mixed-flow pump with axial diffuser.

stator parts (inlet, discharge parts, or spirals), it may be more advantageous to use unstructured or hybrid computational meshes.

For regions close to walls, it is necessary to create finer cells to correctly model the flow in the boundary layer. It is emphasized that at the computational mesh near the wall where flow is dominated by boundary layer, there should be at least ten nodes in the direction perpendicular to the wall. When automatic wall function

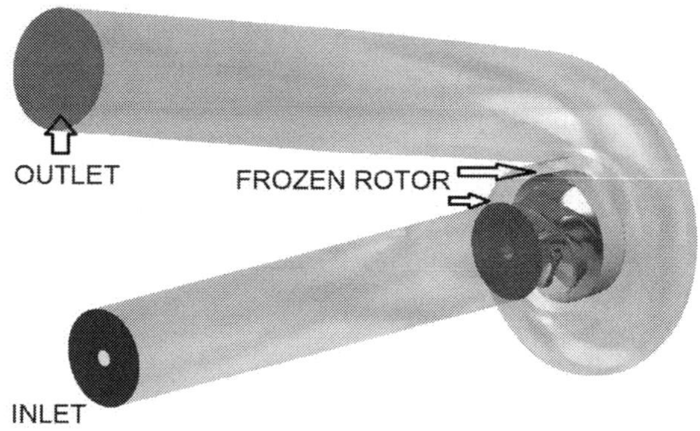

FIGURE 9.8 Computational domain of mixed-flow pump with a spiral casing.

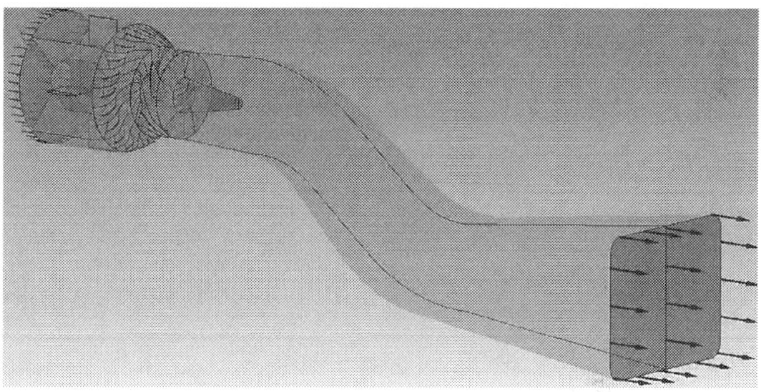

FIGURE 9.9 Computational domain of tubular Kaplan turbine.

is implemented, the recommended value of y+ is maximum of 2. It is advisable to always analyze the effect of mesh density in wall regions and, if possible, compare it to experimental results. If the results are unacceptable, then mesh refinement must be performed.

It is a well-known fact that not only the mesh density in the wall regions but also the total number of elements influences the results of the simulation. It is common practice to perform a mesh sensitivity analysis. The procedure is that first a coarse mesh is used, and the simulation is run. The mesh density is then increased, and the simulation is run; and results are compared. As the density is increased, the model's numerical solution will gravitate towards a unique value. As the model is improved, the amount of computer resources required to execute the simulation also grows. When additional mesh refining causes a minimal change in the solution, the mesh is said to be converged. Gulich recommends the use of 80,000–100,000 elements per passage in between blades. An example of mesh sensitivity analysis for a pump is presented in Table 9.3.

TABLE 9.3
Comparison of Computational Fluid Dynamics (CFD) Simulation Results for a Model Pump $n_b = 0.21$ ($Q_n = 85$ l/s, $Y_n = gH_n$ 140 J/kg, $n_n = 1800$ rpm)

Number of cells	Torque M_k [Nm]	Specific energy Y [J/kg]	Efficiency η [%]	Power P [kW]
89300	73.93	121.07	73.85	13.94
642580	75.95	130.34	77.38	14.32
1272596	76.85	133.98	78.62	14.49
2478408	77.3	135.65	79.14	14.57

Note: The simulation was run at the nominal flow rate, $Q = 85$ l/s.

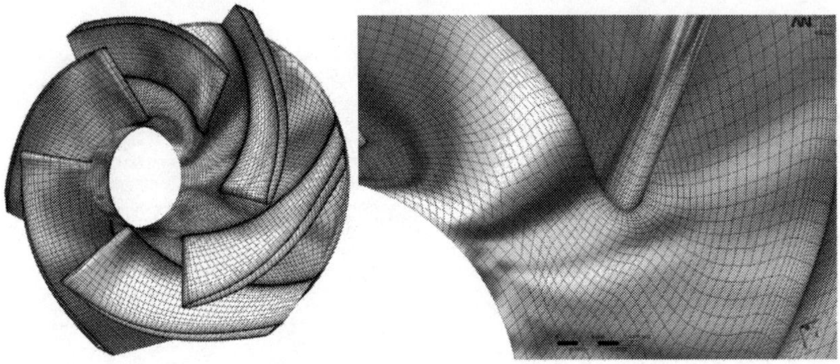

FIGURE 9.10 Structured hexahedral mesh for an impeller.

Based on the results, it can be stated that the number of hexahedral cells – 2–3 million for the whole pump – is sufficient, as given in Table 9.3.

Figure 9.10 shows an impeller with mesh; the results are given in Table 9.3. Detailed view of the mesh near the wall that captures boundary layer flow is given on the right-hand side. The mesh of the impeller and diffuser, including detailed view near the wall region, is given in Figure 9.11.

Figure 9.12 shows an example of mesh creation for a spiral casing, with a detailed view of the wall region. Note that the mesh is denser near the wall and coarser away from the wall region.

Figures 9.13 and 9.14 shows examples of the mesh generation for a tubular Kaplan turbine and a draft tube of a Kaplan turbine, respectively. Note that close to the wall region, the mesh is dense as compared to other regions.

9.5.3 General Settings, Boundary, and Initial Conditions

After creating the computational mesh, it is necessary to specify the preprocessor properties of the considered fluid (fluids in multi-phase model), select specific equations describing flow, and specify simulation methods, interfaces, boundary

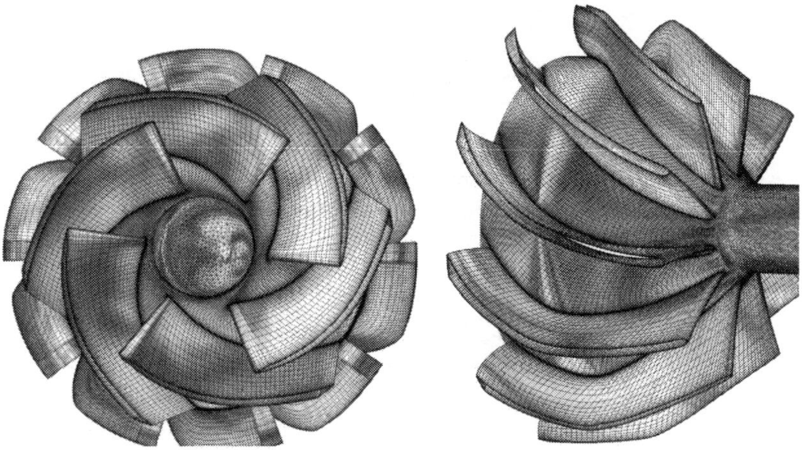

FIGURE 9.11 Structured hexahedral mesh for an impeller and diffuser.

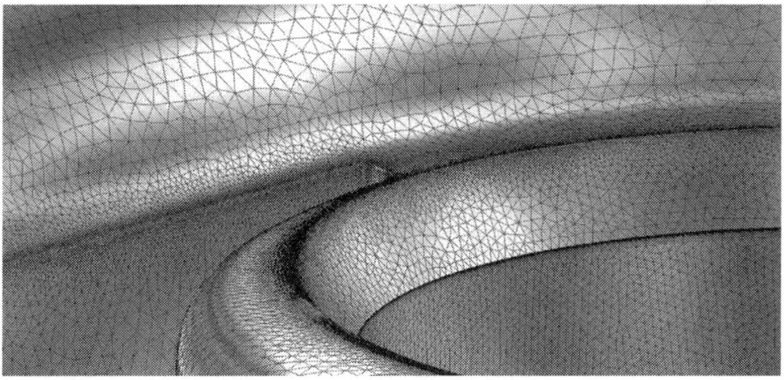

FIGURE 9.12 Detailed view of mesh near the wall of a spiral casing of a pump.

conditions, and initial conditions. In unsteady analysis, it is necessary to specify time step, size, and number of time steps.

Part of the preprocessing work is also the setting of the so-called monitor points, i.e., expressions, by which the desired quantities are monitored and evaluated during the simulation. The desired quantities are shown in the convergence plot at monitor points in the model during the process. They allow for observation of trends and monitor convergence at certain (important) points. When choosing a time step, it is advisable to follow the recommendations given in the manual applied to the CFD software. In general, the size of the time step can be chosen so that, at a given speed, one time step corresponds to a rotation of the rotor by about 2°. The literature recommends performing simulation for different time steps. In general, it is necessary to simulate at least as many time steps as are necessary to obtain an initial periodically

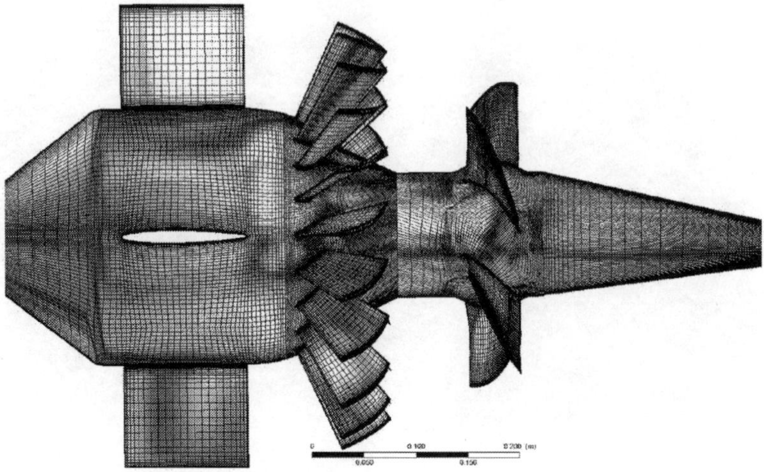

FIGURE 9.13 Mesh of a tubular Kaplan turbine.

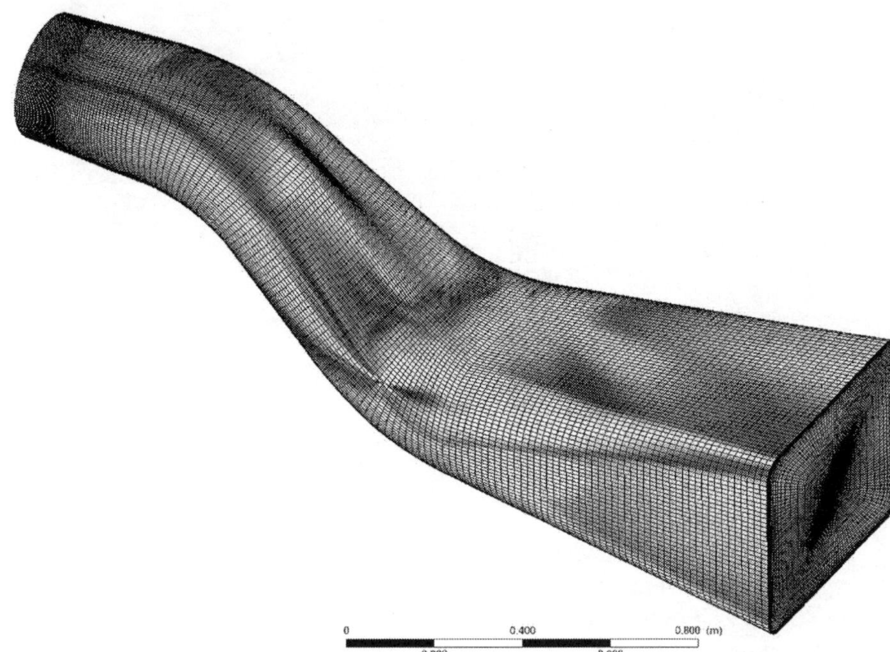

FIGURE 9.14 Mesh of an S – shaped draft tube for Kaplan turbine.

repeating solution. Especially at flow rates lower than optimal, this can be up to about 2000 time steps.

The initial conditions for the stationary analysis may be zero velocity and a static pressure corresponding to the pressure defined as the boundary condition at the inlet or outlet. In the case of a transient analysis, the result of the stationary analysis for a given speed and flow rate may be used as the initial flow field.

Commonly applied boundary conditions are:

Inlet:

- mass flow rate or velocity (either constant throughout the inlet zone or specific velocity distribution). As a rule, it is sufficient to apply a constant speed throughout the cross-section
- turbulence intensity
- hydraulic diameter of the inlet zone.

In some cases, it is more advantageous to apply the following boundary conditions:

- total pressure
- turbulence intensity
- hydraulic diameter of the inlet zone.

Wall zone:

- zero relative velocity
- roughness (the value must not be greater than the thickness of the first layer of cells). New smooth walls can be analyzed using zero roughness
- if it is a stationary wall forming the boundary of the rotating part of the computational domain (the wall of the stator body when modeling a narrow gap above the tip of the open impeller blade), the setting "counter-rotating" wall applies.

Outlet:

- static pressure (where the total pressure is defined at the inlet, the mass flow rate at the outlet must be defined instead of the static pressure)
- turbulence intensity for possible backflow
- hydraulic diameter of the outlet zone.

9.5.4 Simulation

During the simulation, the general convergence criteria (residue curves) must be monitored, while the simulation should be stable, and the residual values should reach sufficiently small values at the end (usually given in the CFD software manual). During the simulation, there may be so-called pseudo instability. Globally, however, the calculation should be stable, i.e., the values of residues should gradually decrease, or remain at a constant value. In addition to the residuals, it is appropriate to monitor the ongoing values of the monitored output parameters such as head, H, and efficiency, η, which should not change fundamentally during the last few tens of iterations (at the end of the simulation).

9.5.5 Model Refinement

The accuracy of the calculation can be improved by applying a more complex model (e.g., unsteady analysis instead of steady, more time steps), better suitable turbulence model, better-suited type of rotor–stator interface, improved quality and density of computational mesh, and more precise specification of boundary conditions. Attention must also be paid to specifying the inlet boundary condition (if the exact velocity profile in front of the machine is not known, the inlet zone must be placed far enough away from the machine to minimize any error due to idealization of the inlet boundary condition). In some cases, it is very important to include some details in the simulation, e.g., narrow gaps between the tip of the blades and the impeller chamber. If the requirement is to determine volumetric losses in detail, the simulation should include sealing gaps as well. If there is a need to determine the axial force, the space behind the rear disc and the space in front of the front disc of the impeller in the simulation should also be included.

9.6 PROCESSING CFD SIMULATION RESULTS

The primary results of CFD simulation are the distribution of velocity and pressure and other quantities in the internal space of the analyzed machine (values of individual quantities usually relate either to the nodal points of the computational mesh or to the centroids of the computational cells). From these results, the integral values of the monitored results are calculated using the functions specified in section 9.6.2, below. Examples of graphical representations of CFD simulation results are shown in Figures 9.15–9.18.

9.6.1 Checking the Accuracy of the Results

After completing the simulation, the total mass and energy flow balances between the inlet and outlet should be checked. The differences in mass and energy flow between inlet and outlet should be less than 0.3%. Another important parameter is the value of y^+, which should be on all solid walls in the range that corresponds to the applied wall function. At the same time, it is necessary to perform a global check of the distribution of individual quantities of interest in the flow field, which must be physically meaningful.

9.6.2 Evaluation of Monitored Parameters

Pump specific energy, pump head multiplied by acceleration due to gravity ($Y_p = Hg$) is the difference of specific energies at the outlet and at inlet of the pump. This can be written as Equation (9.62).

$$Y_p = \frac{p_{TotOut} - p_{TotIn}}{\rho} \tag{9.62}$$

where

ρ is fluid density

p_{TotOut} is outlet total pressure determined as the mean integral value at the outlet

p_{TotIn} is inlet total pressure determined as the mean integral value at the inlet.

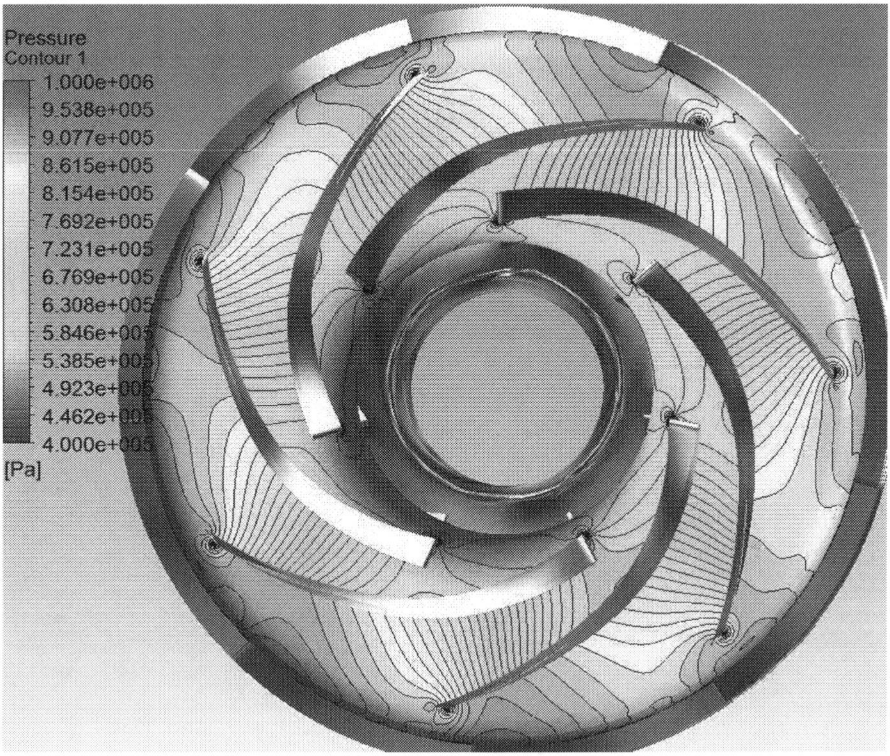

FIGURE 9.15 Pressure distribution in between impeller blades.

FIGURE 9.16 Pressure distribution in the channels of the impeller and diffuser.

FIGURE 9.17 Pressure distribution in the channels of the impeller and the spiral casing.

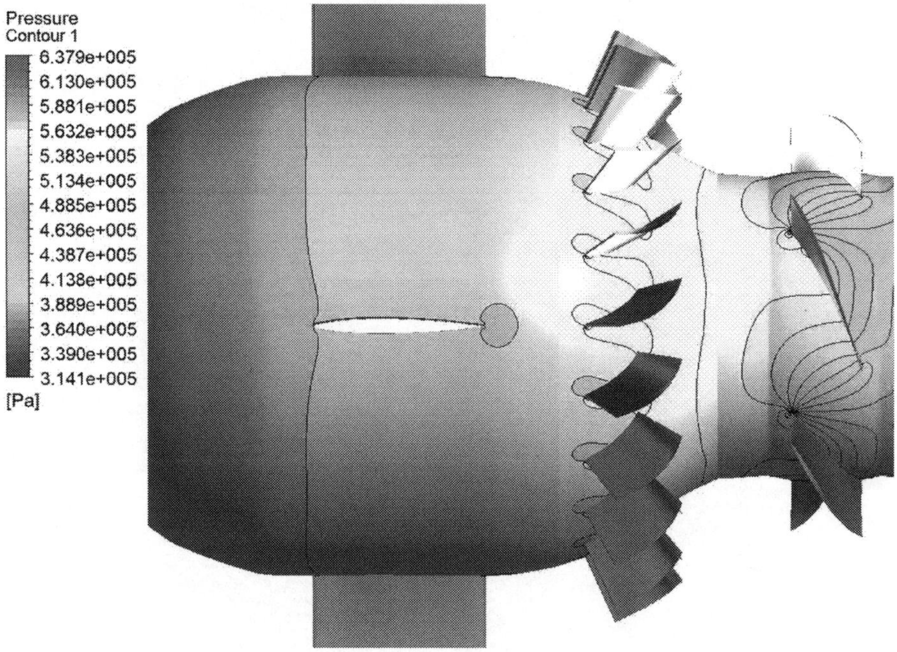

FIGURE 9.18 Pressure distribution in the channels of the impeller, distributor, and in the inlet of the Kaplan turbine.

The total pressure is given by the following equation:

$$p_{Tot} = \frac{\sum \dot{m}_{(i)} \cdot p_{Tot(i)}}{\sum \dot{m}_{(i)}} \tag{9.63}$$

where

$\dot{m}_{(i)}$ is local mass flow rate through the wall of the relevant cell
$p_{Tot(i)}$ is the local value of the total pressure corresponding to the considered cell.
Equation (9.64) can be used to compute the total pressure in the considered cell.

$$p_{Tot(i)} = \frac{p_{Stat(i)}}{\rho} + \frac{v_{(i)}^2}{2} \tag{9.64}$$

where

$p_{Stat(i)}$ is static pressure of a given cell
$v_{(i)}$ is absolute speed corresponding to the cell under consideration
$p_{Tot(i)}$ is the local value of the total pressure corresponding to the considered cell.

The total specific energy of a turbine is also determined in a similar manner using Equation (9.65). The subscript T stands for the turbine.

$$Y_T = \frac{p_{TotIn} - p_{TotOut}}{\rho} \tag{9.65}$$

The head of a pump and a turbine are determined with Equation (9.66). It should be noted that the actual net head of the turbine will be greater by a value proportional to the losses in the penstock.

$$\mathrm{H}_{p(T)} = \frac{Y_{p(T)}}{g} \tag{9.66}$$

Equation (9.67) can be used to calculate the useful power of a pump. The useful power of a turbine is determined using Equation (9.68).

$$P_{Up} = \rho Q Y_p = \rho Q g H_p \tag{9.67}$$

$$P_{UT} = M_k \omega = M_k 2\pi n \tag{9.68}$$

where

Q is flow defined as boundary condition
P_{up} is useful pump power
M_k is torque due to the action of fluid on the rotor.

The resulting torque to be considered is the sum of the torques acting on the impeller blades and discs and on other rotating surfaces associated with the impeller (e.g., shaft).

ω is the angular velocity of the rotor
n is rotation speed of the rotor (s^{-1}).

The pump power P_p is calculated using Equation (9.69).

$$P_p = M_k \omega = M_k 2\pi n \tag{9.69}$$

The theoretical power delivered to the turbine can be expressed according to Equation (9.70), which is slightly higher than the shaft power.

$$P_T = \rho Q Y_T = \rho Q g H_T \tag{9.70}$$

Equation (9.71) can be used to determine the hydraulic efficiency of the pump, η_{hp}, and the hydraulic efficiency of the turbine, η_{hT}.

$$\eta_{hP(T)} = \frac{P_{UP(UT)}}{P_{P(T)}} \tag{9.71}$$

The determination of axial and radial forces (F_{AX}, F_{RAD}) on solid surfaces can usually be performed using functions programmed directly in the postprocessor of the simulation software. The determination of the net positive suction head (NPSH) of the pump is usually performed on the basis of repeated CFD calculations with varying static pressure on the suction (decreasing the pressure gradually – the pressure must therefore be defined as the inlet boundary condition). The NPSH value then corresponds to the NPSH on the pump suction for the calculation at which the calculated specific energy falls below a predetermined value (e.g., 3% below the original specific energy or head value).

9.7 EXAMPLE PROBLEMS

Example 9.1

Water flows through the fitting as shown in Figure 9.20. The inlet is on the left and the fitting branches out as shown, and the outlet is through both branches. The inlet and outlet branches are each 1 m long. The angle between the axis of the first (also the second) output branch and the inlet branch is 30° (the axes of the output branches form an angle of 60° with each other). The inner diameter of the inlet and outlet branches is 200 mm. The inner walls of the fitting are hydraulically smooth. The velocity of water flow at the inlet is 7.5 m/s and there is atmospheric pressure at the outlet of both branches. Calculate the resulting total force from the flowing water on

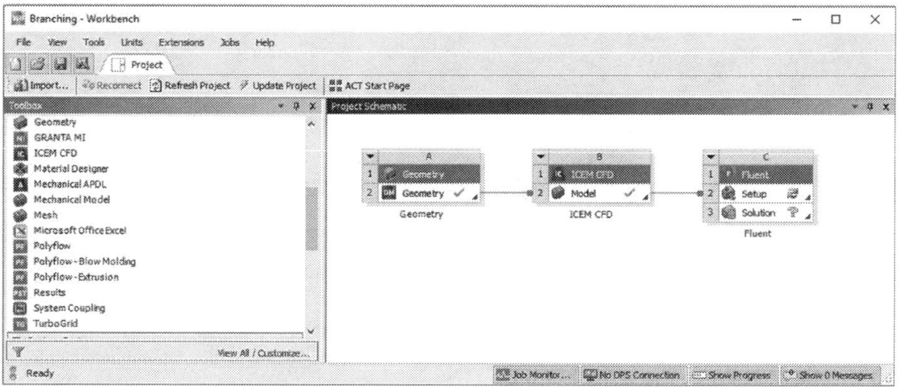

FIGURE 9.19 ANSYS Workbench graphical interface.

the inner surface of the fitting through which the water flows. Consider clean water at 15°C. The entrance to the fitting is connected to a long straight pipe with a length of 150 D. Neglect the influence of gravity.

There are three ways to solve the problem. The first method is CFD solution with consideration of viscous flow (the influence of real water viscosity is considered); the second method is a CFD solution neglecting the effect of water viscosity (nonviscous flow); and the third method is an analytical solution with consideration of 1D flow of ideal liquid (neglecting the influence of viscosity). Compare the results.

Solution

It is possible to use your own CFD program, free CFD software (freeware or open-source software such as OpenFOAM), or commercial CFD software to solve the problem using CFD methods. In this case, ANSYS commercial software is used.

Individual parts of CFD simulation (creation of geometry, creation of computational mesh, definition of boundary conditions, simulation, and postprocessing) are implemented in separate software modules of the ANSYS software. Data input (files) for subsequent modules can be imported into the respective software modules "manually" via commands available in the graphical interface of the respective modules. The second option is to integrate individual modules used in CFD simulation through the so-called ANSYS Workbench, as shown in Figure 9.19. At the same time, a file representing the CFD simulation project and file folder, containing all files of the given CFD simulation (files with geometry of the computational mesh settings, etc.), is created. The flow of data (file transfer) from one module to the next (for example, transfer of a geometry file to a meshing module) is defined by "dragging and dropping" by left clicking on the block diagram of the modeler and dragging the cursor to the block diagram of the meshing module.

The geometry is created in the ANSYS Design Modeler program (Figure 9.20). The geometry in principle comprises the inner space of the fitting, i.e., the volume in which the water flows. The fixed walls of the fitting themselves are not part of the simulation. Smaller shape details (for example, curvatures at the joint of individual

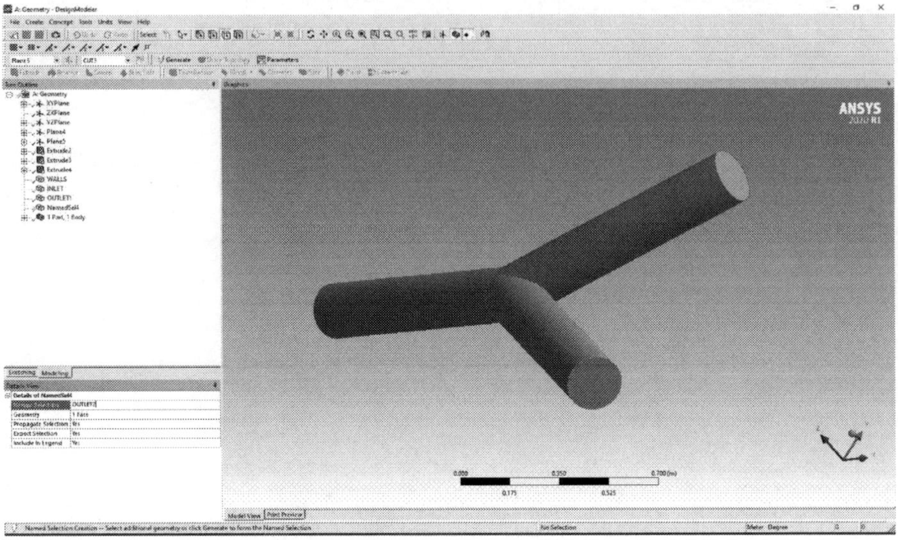

FIGURE 9.20 Geometry of the fitting.

branches) can be neglected. Alternatively, the geometry can be created in any other 3D computer-aided design (CAD) software and imported into the ANSYS environment using one of the exchange CAD file formats (e.g., STP). After defining the geometry of the fitting, it is necessary to define the individual boundary zones on which the boundary conditions are defined. In the ANSYS Design Modeler environment, the boundary zones are defined using the so-called "Named Selection" of objects (see the tree of the finished 3D model in Figure 9.20). The boundary zones are: INLET – entrance to the computational domain, OUTLET1 and OUTLET 2 – outlet from the computational domain, WALLS – representing the inner surface of the fixed walls of the fitting (the fixed walls themselves are not part of the simulation).

A computational mesh can be created in the ANSYS software environment using the ANSYS Mesh, ICEM, or fluent meshing modules. In this case, the ANSYS ICEM software module was used.

Among the available types of meshes are:

- Hybrid mesh – consists of prismatic cells near solid walls (boundary layer area) and quadrilateral cells in the area of developed flow. The transition between prismatic cells and quadrilaterals is realized by pyramid computational cells.
- Hybrid mesh – consists of prismatic cells near solid walls (boundary layer area) and hexahedral cells in the area of developed flow. Both parts of the mesh are automatically connected by pyramid computational cells.
- Polyhedral mesh – the main advantage of a polyhedral mesh is that each individual cell has a large number of neighbors, allowing gradients to be accurately simulated. Polyhedrons are less susceptible to stretching than tetrahedrons. This results in enhanced mesh quality and model numerical stability.
- Hexahedral mesh – can be structured or unstructured.

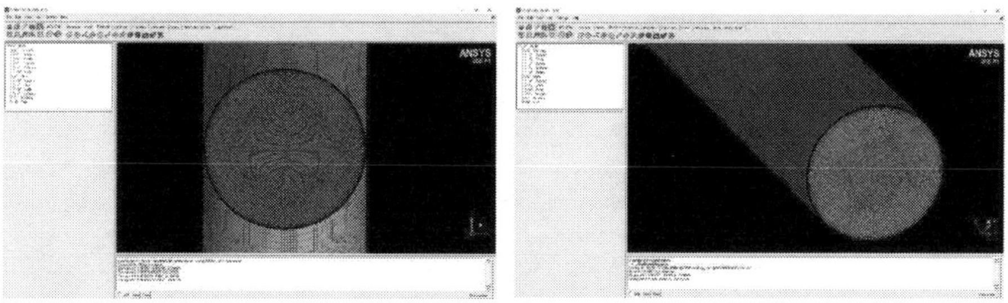

FIGURE 9.21 Inlet and outlet mesh.

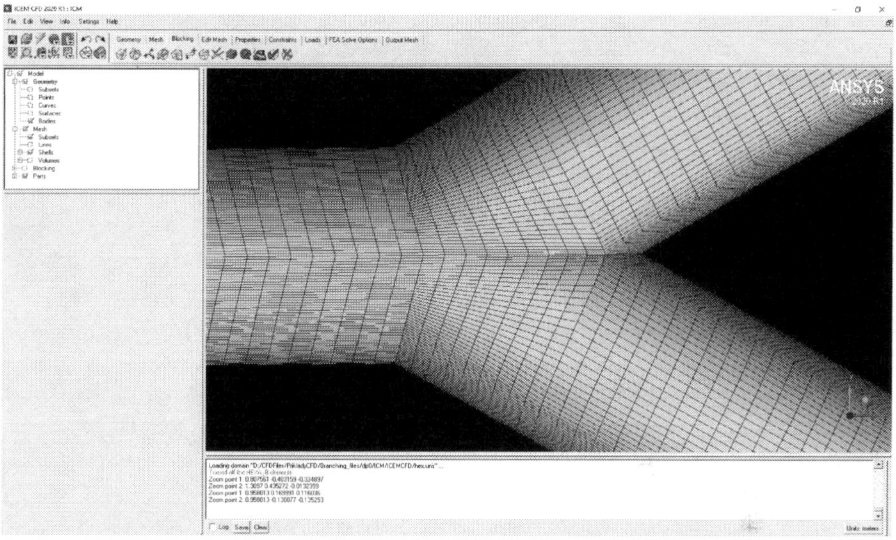

FIGURE 9.22 Detailed view of the mesh at the branching point of the fitting.

Details of the individual options for creating a computation mesh are available in the manuals of specific software. For our example, a structured hexahedral mesh, shown in Figures 9.21 and 9.22, was used. When creating a mesh, it is necessary to ensure that the growth rate of computed cells – growth rate (ratio of lengths of two adjacent cells) does not exceed 1.2. Furthermore, the aspect ratio (length to width ratio) should not be greater than 100 (except for the boundary layer, where it may exceed 1000). In addition, it is necessary to monitor the values of the skewness, which should not exceed 0.95. The qualitative parameters of the mesh can be evaluated directly in the software in which the mesh is created.

If a wall function is used, the thickness of the first layer of cells must be chosen so that after performing the CFD simulation the values of dimensionless distance from the wall (y^+) are reached in the range required. If wall function is not used the thickness of the cell is chosen so that the value of y^+ is usually less than 2.

The number of computational cells (elements) must be selected on the basis of the analysis of the impact of computational mesh density (mesh independence study).

This is also called mesh sensitivity analysis. The goal of the mesh independence study is to eliminate or minimize the impact of the number of cells on the computational results. And it is always a good idea to do it for each unique geometry, which may be laborious. The mesh independence study for a single geometry will only apply to that geometry. Typically, a series of coarse, medium, and fine meshes is created to demonstrate that the solution does not vary much between successive iterations.

The next step of the preprocessing is mesh importing into the environment of the fluent simulation module. In the ANSYS Workbench work environment (Figure 9.19), it is sufficient to ensure dataflow between the mesh generation module and the fluent simulation module (Figure 9.19). Subsequently, model specifications are applied, such as the selection of a steady or unsteady flow model. In this example, it is sufficient to consider steady flow. In the next step, the turbulence model is set up (since the flow is turbulent, as evidenced by the high value of the Reynolds number). For the purposes of basic engineering calculations such as this one, it is sufficient to apply the k-ω SST model, which is robust (ensures good stability and convergence of the solution) and is able to relatively accurately model flow with possible boundary layer separation. When applying the SST model, it is not necessary to implement the selection of the wall function separately. This is because the SST model implements specific formulations of the wall function/procedures for dealing with flow around walls. If we are neglecting the influence of viscosity, a nonviscous flow model is defined in the settings instead of the turbulence model.

Next, the physical properties of the flowing liquid are defined. It is necessary to specify the exact value of density and viscosity (in viscous flow) corresponding to the temperature of the liquid. Water with a temperature of 15°C has a density of 999 kg/m^3 and its viscosity is 1.1456×10^{-6} m^2/s. When defining viscosity, it is necessary to distinguish whether the software works with kinematic or dynamic viscosity. After defining the physical properties of water, it is necessary to define water as a "liquid."

Next, we set the boundary conditions. The outlet pressure specifies the outlet pressure values, therefore both outlet boundary zones are defined as "pressure outlet"-type zones. In both cases, the same settings apply:

- average pressure value (constant value for the whole zone)
- turbulence intensity (5%, which is the normal value of turbulence intensity in a circular pipe)
- hydraulic diameter of the outlet zone (in this case it is 0.2 m).

Fixed walls are defined as "wall"-type zones, with a zero-roughness value in the settings and at the same time considering a zero-flow velocity over the entire wall.

The inlet boundary condition can be defined in several ways. In this case the average value of the velocity at the inlet was used. This velocity can be defined as a constant value for the entire inlet zone. It is also possible to use the mass flow rate, which is calculated using the area of the inlet, mean velocity, and density. The mass flow rate can then be applied as a boundary condition instead of the velocity. However, both methods ignore the real velocity profile at the entrance. Therefore, for greater accuracy it may be better to determine the velocity distribution at the inlet (velocity profile) and define it via a file as a boundary condition. The velocity profile file itself can be specified as follows.

A separate CFD analysis of the flow in a straight pipe will be performed, the outlet boundary zone of which is identical in its geometry and location in space with the entrance to the solved fitting. The length of the straight pipe must be at least 40 D. After performing the simulation, the velocity distribution at the outlet is exported using the appropriate function in the fluent software. This will create a separate file with a specific velocity profile. When defining this profile as a boundary condition at the inlet of the fitting, it is first necessary to import this file also in the fluent software environment and then apply the velocity component setting in the x, y, z-axes when setting the boundary condition. The hydraulic diameter of the inlet boundary zone (0.2 m) and the turbulence intensity (5%) must also be set at the inlet.

The simulation is started by "initializing" it. For this example, the course of the simulation should be stable, and the solution should converge within 150–200 iterations. The convergence criteria are the residuals of the monitored quantities, which should reach the required value. In addition, it is necessary to define and monitor the "monitor points," i.e., expressions that evaluate the sizes of selected quantities in specific places of the computational domain. One of the "monitor points" is the force acting on the fitting, which we defined via the menu item of the fluent software called "monitors." The magnitude of the force is defined using the built-in function of the fluent software, which evaluates the forces acting on selected boundary zones of the "wall" type. The value of "monitor points" can be monitored during the simulation by plotting the values in the form of a graph in the graphical user interface (GUI), while the x-axes are individual iterations, and the y-axes are the values of "monitor points." It is also possible to set the writing of "monitor point" values during iterations to a file. The purpose of monitoring monitor points during the simulation is to achieve such a state that with increasing iterations the value of monitor points change is negligible. Then it is possible to consider the solution as converged. If the residuals reach the required values and the values of the "monitor points" still change significantly, it is necessary to reduce (for example by one order) the required residual values and continue the simulation. Other "monitor points" can be the mean values of the inlet pressure and outlet velocity.

After the simulation is complete, it is necessary to check the mass flow balances between the inlet and outlet. The difference between the incoming and the sum of the exiting mass flows must not be greater than 0.3% of the value at the inlet. Furthermore, it is necessary to make sure of the physical meaningfulness of the calculated pressure and velocity distribution (or other quantities of the flow field). One of the possibilities is to evaluate and graphically represent and qualitatively analyze the distribution of pressure and/or velocity in the plane passing through the axis of the fitting, as shown in Figure 9.23. The values of the quantities must change continuously; physically unreasonably high gradients must not be present. In places of unreasonably high pressure or velocity gradients, it is possible to carry out local refinement of the computational mesh and perform the simulation again.

The calculated total force on the fitting is obtained either from the "monitor point" or it can be determined using the calculator implemented in the fluent software.

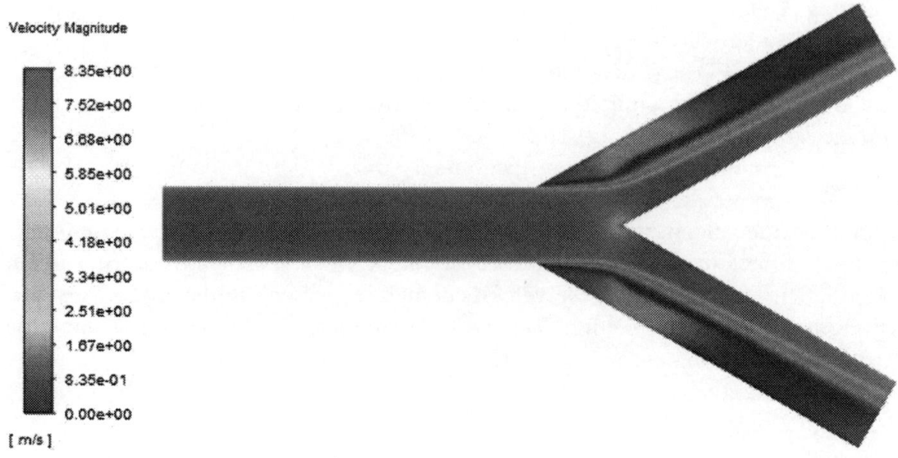

FIGURE 9.23 Velocity distribution in the fitting.

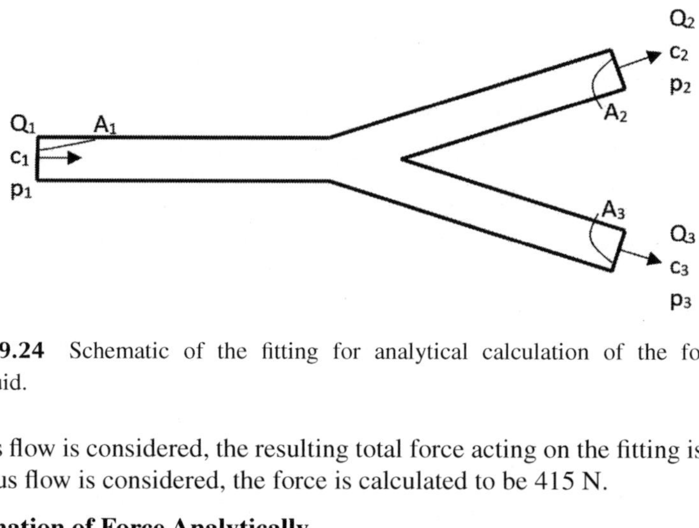

FIGURE 9.24 Schematic of the fitting for analytical calculation of the force of the flowing fluid.

If viscous flow is considered, the resulting total force acting on the fitting is 585 N. If nonviscous flow is considered, the force is calculated to be 415 N.

Determination of Force Analytically

We assume a 1D, nonviscous flow. The description of the quantities at the inlet and outlets of the fitting is shown in Figure 9.24. Note that c_1, c_2, and c_3 are inlet and outlet velocities, respectively.

If we neglect the gravitational force, Equation (9.72) can be used to calculate the resultant force acting on the fitting.

$$F = \sqrt{F_x^2 + F_y^2} \tag{9.72}$$

$$F_x = F_{p1x} + F_{p2x} + F_{p3x} + \rho Q_1 c_1 - \rho Q_2 c_2 \cos(30°) - \rho Q_3 c_3 \cos(-30°) \tag{9.73}$$

$$F_y = F_{p1y} + F_{p2y} + F_{p3y} - \rho Q_2 c_2 \sin(30°) - \rho Q_3 c_3 \sin(-30°) \tag{9.74}$$

$$F_{p1x} = p_1 A_1 \tag{9.75}$$

$$F_{p2x} = p_2 A_2 \cos(30°) \tag{9.76}$$

$$F_{p3x} = p_3 A_3 \cos(-30°) \tag{9.77}$$

We recognize that gage pressures p_2 and p_3 are equal to zero. Hence the forces arising from these pressures are also equal to zero. Similarly, for the *y*-axis direction, the forces F_{p2y}, F_{p3y} are equal to zero. Also the force F_{p1y} is equal to zero. From Bernoulli's equation it is possible to derive Equations (9.78) and (9.79).

$$p_1 = p_2 + \rho\frac{c_2^2}{2} - \rho\frac{c_1^2}{2} \tag{9.78}$$

$$p_1 = p_3 + \rho\frac{c_3^2}{2} - \rho\frac{c_1^2}{2} \tag{9.79}$$

From Equations (9.78) and (9.79), it is clear that the velocity c_2 is equal to the velocity c_3. We also recognize the equation of continuity, Equation (9.80).

$$Q_1 = Q_2 + Q_3 \tag{9.80}$$

$$Q_1 = c_1 A_1 \tag{9.81}$$

$$Q_2 = c_2 A_2 \tag{9.82}$$

$$Q_3 = c_3 A_3 \tag{9.83}$$

$$A_1 = \frac{\pi D_1^2}{4} \tag{9.84}$$

$$A_2 = \frac{\pi D_2^2}{4} \tag{9.85}$$

$$A_3 = \frac{\pi D_3^2}{4} \tag{9.86}$$

Substituting (9.81)–(9.86) into Equations (9.80) and considering the equality of both output velocities, we get Equation (9.87).

$$c_2 = c_3 = \frac{c_1 A_1}{A_2 A_3} \tag{9.87}$$

Substituting Equation (9.87) into Equation (9.74), we see that the resulting force in the direction of the y-axis is equal to zero. After substituting numerical values into Equations (9.72) and (9.73) we get a resultant force equal to 338.93 N. The difference between the value of force determined by CFD simulation for nonviscous flow and the value of force determined analytically with the assumption of nonviscous fluid, is 76 N.

Example 9.2

Determine the head and hydraulic efficiency of the radial pump impeller shown in Figure 9.25. The pump pumps clean water at a temperature of 25°C. Consider 500 kPa pressure at the impeller outlet. Consider speed $n = 1450$ rpm and flow $Q = 175.2$ l/s.

Solution

The impeller geometry (Figure 9.25) was created in an external CAD modeler. CFD analysis was performed in the ANSYS software environment. The impeller geometry is imported into the ANSYS software environment via the ANSYS Design Modeler. Subsequently, a volume is defined representing the space around one impeller blade (Figure 9.26), which also forms the computational domain. The stated volume must have an adequate overlap before entering the impeller and at the exit. This is clear from the Figure 9.26.

The impeller has seven blades. For this reason, the hydraulic volume needs to be exactly one-seventh of the total internal volume of the impeller. The geometry of the two peripheral surfaces of the volume (passing approximately through the center line of the inter-blade space) must be symmetrical about the pump axis, so that periodicity can be applied in the creation of the computational mesh as well as in the simulation and subsequent evaluation of results. Then it is possible to define the boundary

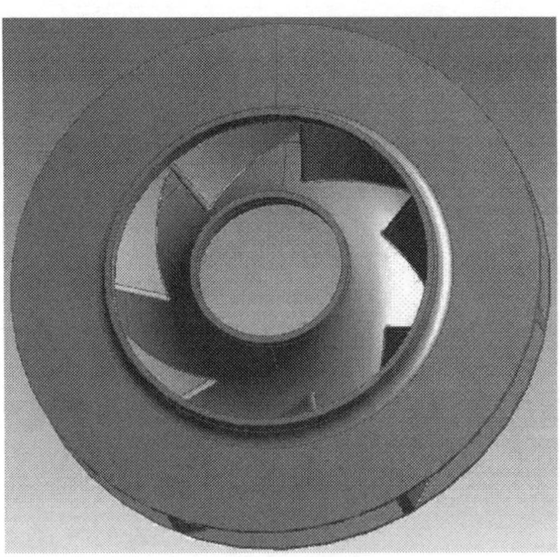

FIGURE 9.25 Radial pump impeller considered.

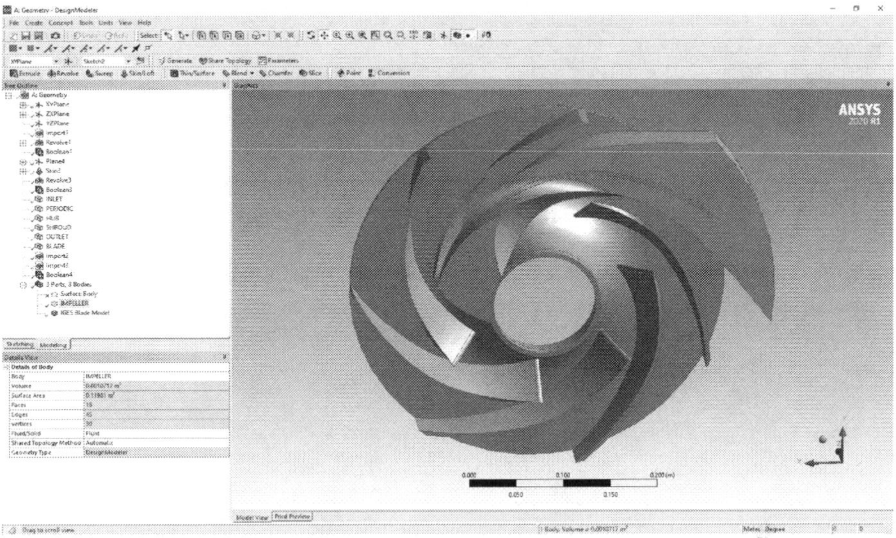

FIGURE 9.26 Volume around the impeller blade for computation.

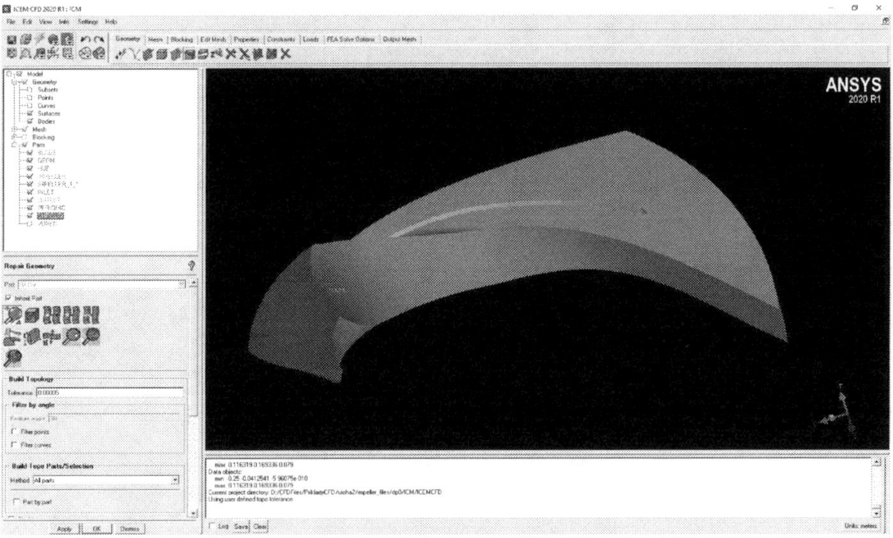

FIGURE 9.27 Geometry of the simulated domain – the space through which the pumped liquid flows around one blade.

zones: INLET (inlet of the computational domain), PERIODIC (side surfaces), OUTLET (outlet of the computational domain), HUB (rear disc or hub), BLADE (all surfaces on the blade surface), and SHROUD (front disc). The HUB, BLADE, and SHROUD zones are zones on which it is then necessary to define wall "zone" zones in the preprocessor (Figure 9.27).

The computational mesh can be created in the ANSYS ICEM CFD software environment as in Example 9.1.

When applying the ICEM software, it is first necessary to define the "periodicity" of the computational domain. This is done via the menu item "Mesh – Global mesh parameters – Set up periodicity." The unit vector of the axis of rotation and the angle corresponding to the section of the computational domain are defined.

The next step is to create blocks. The first step is to initiate the main block (hexagon), which roughly describes the entire computational domain. Subsequently, the individual edges of the block are associated with the respective edges delimiting the computational domain. It is also necessary to define mutually periodic vertices (Blocking – Edit block – Periodic vertices) (Figure 9.28). It is necessary to use the cursor to select a specific vertex and its periodic counterpart, which is on the opposite side of the respective edge (on the opposite surface).

Next, the main block is decomposed into several smaller blocks and their edges are associated with the edges delimiting the individual boundary zones of the computational domain (Figure 9.29). Subsequently, the number of nodes and their layout must be defined on the individual edges of specific blocks. The final meshed surface is shown in Figure 9.30.

The generation of computational mesh must respect the generally applicable principles as described in section 9.5.2. In this case, the thickness of the first layer of computational cells was 10 micrometers, the number of layers between the front and rear discs was 70, the number of layers into which the inlet and outlet edges were divided was 20, the number of layers between the blade side wall and the periodic boundary zone was 16, and the number of layers between inlet and outlet was 60. The proposed computational mesh must also be subjected to sensitivity analysis to ensure accuracy.

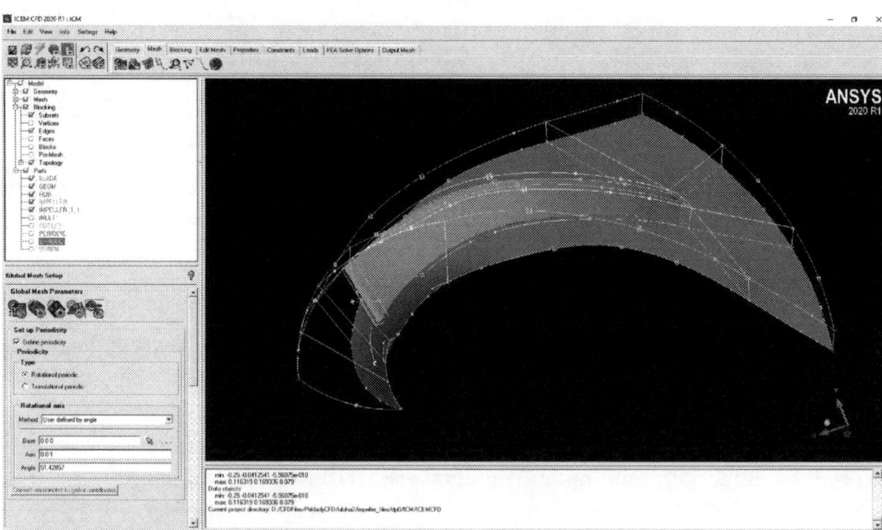

FIGURE 9.28 Computational block boundaries and method of defining "periodicity".

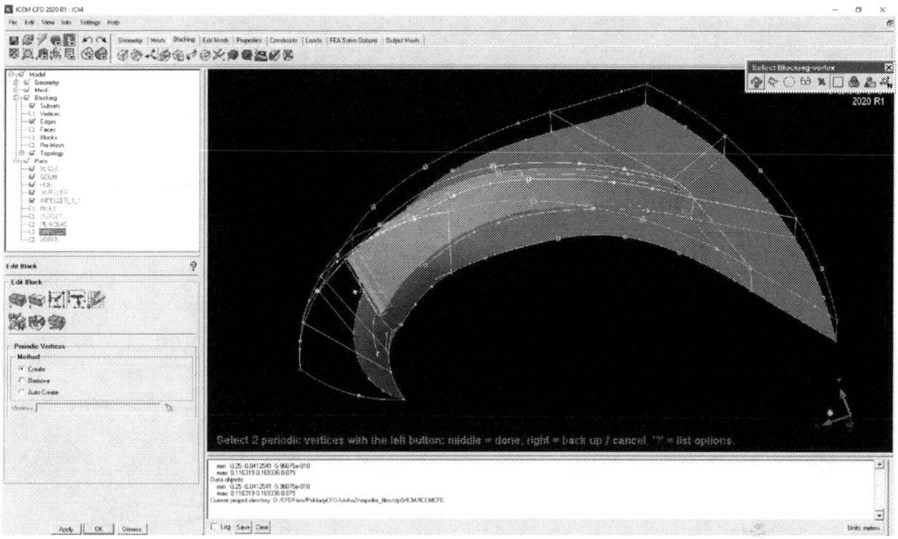

FIGURE 9.29 Block boundaries and method of defining mutually periodic individual vertices, edges, and walls of blocks.

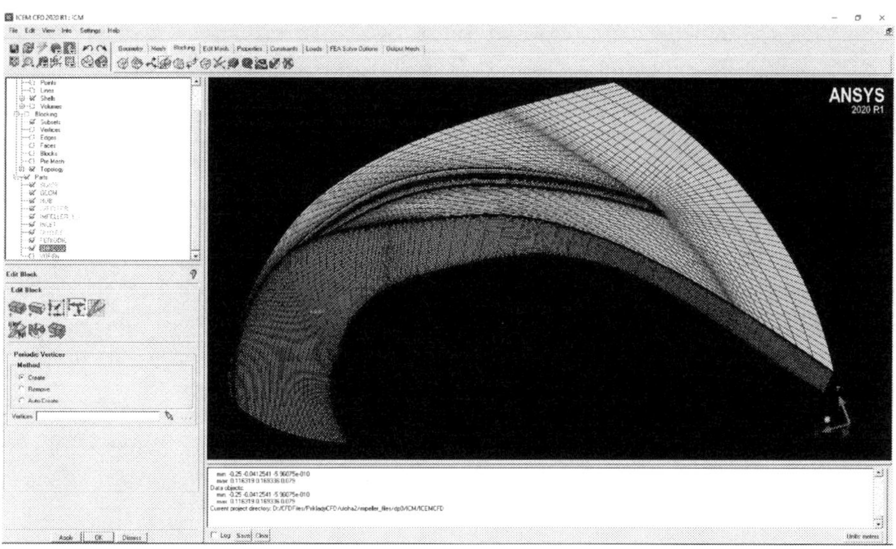

FIGURE 9.30 Meshed surface.

After creating the computational mesh, the CFD simulation settings are selected in the ANSYS CFX Pre environment. In the preprocessor, it is advantageous to use the built-in "turbo wizard" in the calculation settings. When applying the turbo wizard, it is necessary to gradually specify the following data concerning the impeller. In the introductory window the type of turbomachine is specified; in this case it is a hydrodynamic pump ("pump"). At the same time, the type of analysis is specified – steady analysis ("steady state"). Subsequently, it is necessary to specify the boundary zones, and the direction of rotation of the impeller. Since the boundary zone names have been defined, specific zones are automatically recognized, and a specific type of boundary condition is assigned. SHROUD, WALLS, HUB, BLADES zones are automatically recognized and defined as wall-type zones. The INLET zone is recognized and defined as inlet and the OUTLET zone as outlet. PERIODIC zones should be defined as periodic boundary zones. Subsequently, the main settings of the model are determined. In this case, the SST turbulence model is specified. The advantages of applying the SST model are explained in more detail in section 9.1.4.1.5.3. Furthermore, the second-order discretization schemes need to be set – in the CFX environment these are referred to as "high resolution" – and the type of fluid under consideration, in this case water. In addition, the size of the time scale is determined. It may be advantageous to choose the "physical time scale" option, which is a fixed value. Convergence depends to a large extent on this value. Too small a value can cause extremely slow convergence and an inappropriately large value can cause computational instability or divergence. In general, it is advisable to apply the recommendations given in the ANSYS CFX software manual when selecting this parameter. In this case, a value of 0.005 was applied. Furthermore, the specific values of the boundary conditions should be confirmed. We define the inlet as "mass – flow inlet" and specify the value of the mass flow rate belonging to one-seventh of the considered mass flow through the impeller. In this case, the mass flow rate is 25 kg/s. The outlet is defined as "pressure – outlet," specifying the average value of the static pressure at the impeller outlet. The default settings of all "wall"-type zones are retained, and the same applies to "periodic" boundary zones. The settings are then completed via the "wizard."

Next, in the solver settings, which are accessible via the menu in the tree structure on the left side of the screen in the GUI program CFX Pre, selections are made (Figure 9.31). The settings are accessed by double-clicking on the "solver control." In these settings, it is necessary to adjust the applied discretization scheme for the turbulence model equations to a "high-resolution" scheme (second-order scheme). In addition, it is appropriate to reduce the value of the convergence criterion by one order in order to fully stabilize the monitored values (in this case head and hydraulic efficiency) during the simulation.

The monitored values (i.e., head and hydraulic efficiency) must first be defined as "expressions" via the appropriate commands available in the CFX GUI (Figure 9.32). Specific expressions for defining monitored quantities can be created using commands accessible via the context menu. Details are given in the CFX software manual and in the relevant tutorials. The expression for head is given by Equation (9.66). In a similar way, it is necessary to define the expression for the calculation of hydraulic efficiency, given by Equation (9.71).

Before initiating the simulation, it is appropriate to define the "point monitor" for monitoring the values of predefined terms, i.e., head and hydraulic efficiency. The values of the expressions can then be monitored throughout the simulation in the graphical interface of the CFX solver.

Evaluation of the results after the simulation can be done, either directly in the GUI solver or using the CFD Post postprocessor, where the calculated values of

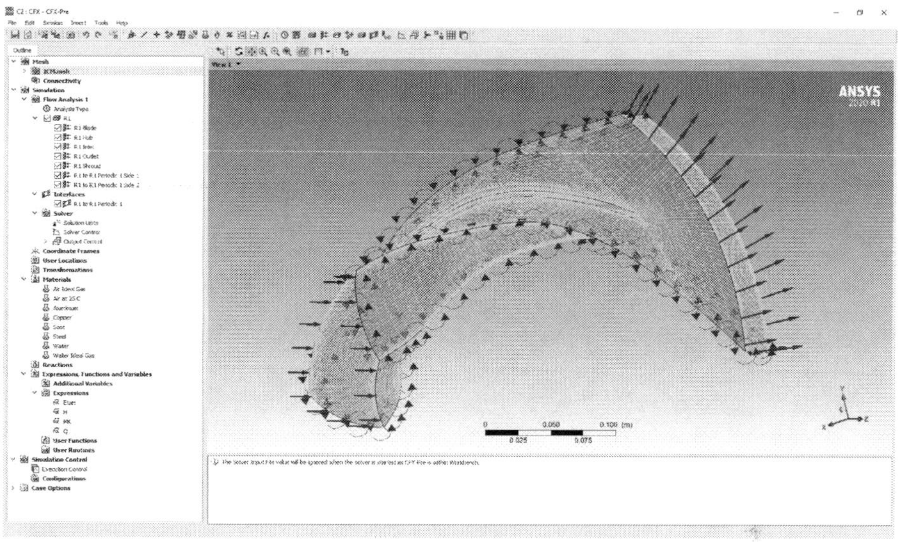

FIGURE 9.31 CFX software preprocessor graphical user interface (GUI).

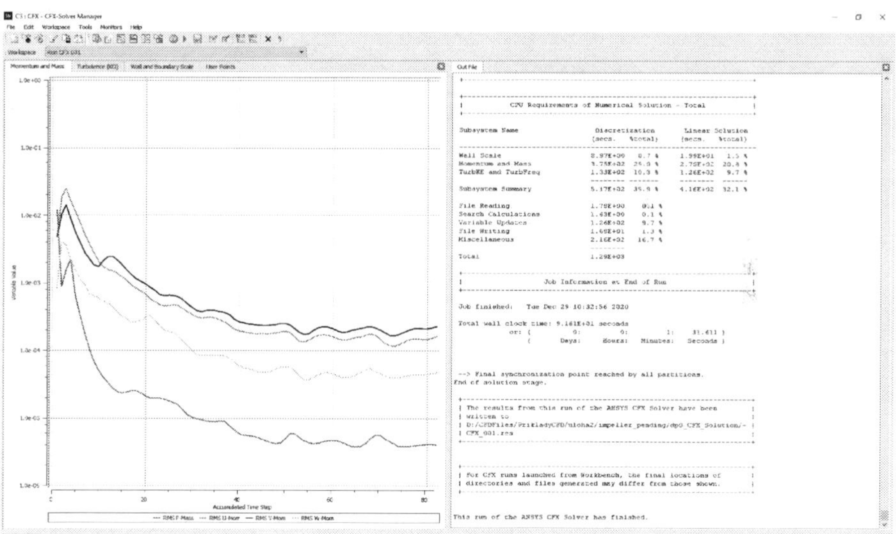

FIGURE 9.32 The course of residuals during the simulation in the graphical interface of the CFX solver.

the expressions for the head and efficiency are available in the item "Expressions." The calculated value of the head in this case is 48.6 m and the hydraulic efficiency is 93.4%. It should be noted that the calculated values correspond only to the impeller itself. The values after considering the losses in the stator parts of the pump are significantly smaller. Also, it should be noted that the calculated hydraulic efficiency does not consider mechanical and volumetric losses.

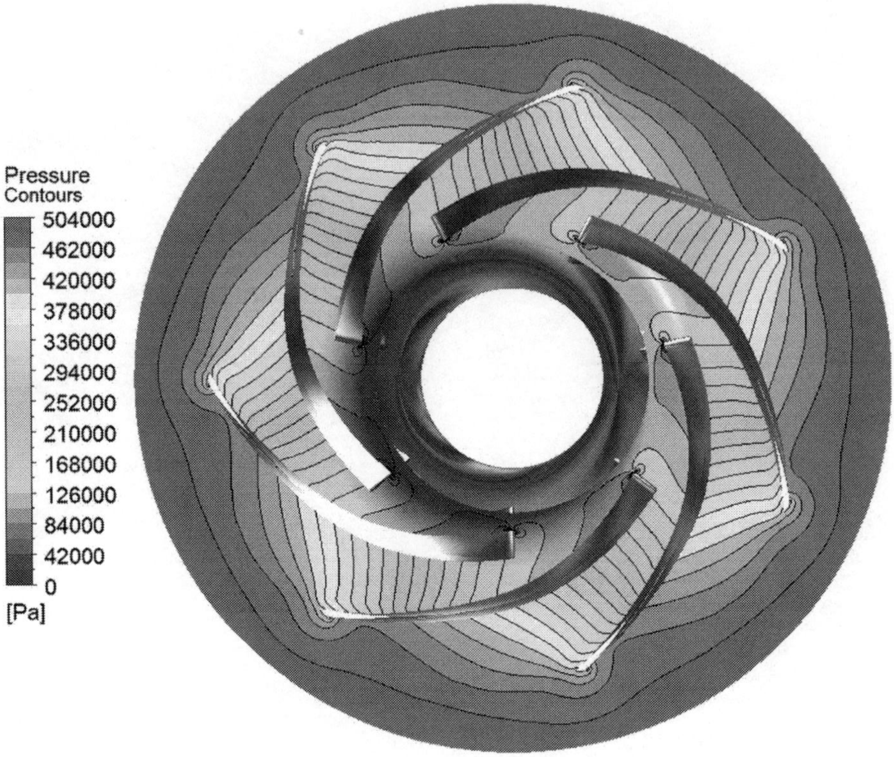

FIGURE 9.33 Static pressure distribution in the impeller. Isobars are shown.

The pressure and velocity distribution can be seen in Figures 9.33 and 9.34, respectively. Displays of monitored quantities for all inter-blade space were created by applying periodicity with respect to the pump axis in the postprocessor.

Example 9.3

The impeller of Example 9.2 (Figure 9.25) works in turbine mode of operation together with a radial stator. The stator blades are shown in Figure 9.35. The stator has 21 blades. Calculate the head and hydraulic efficiency of the impeller coupled with this stator if water with a temperature of 25°C flows through the impeller; the volumetric flow rate is $Q = 385.4$ l/s. Consider a pressure of 300 kPa at the impeller outlet. The rotational speed of the rotor (impeller) is $n = 1450$ rpm.

Solution

The approach to solving this problem is almost identical to that for solving Example 9.2. The differences compared to Example 9.2 are as follows:

- The computational domain includes the inter-blade channels of the stator; therefore, it is necessary to create the geometry of the "fluid volume" around

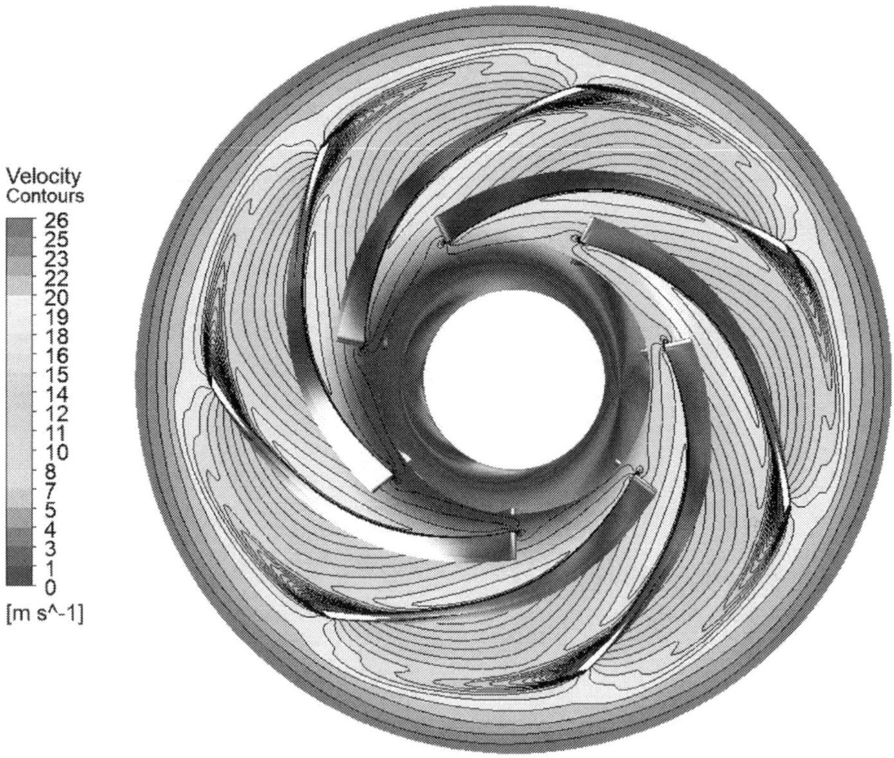

FIGURE 9.34 Velocity distribution in the impeller.

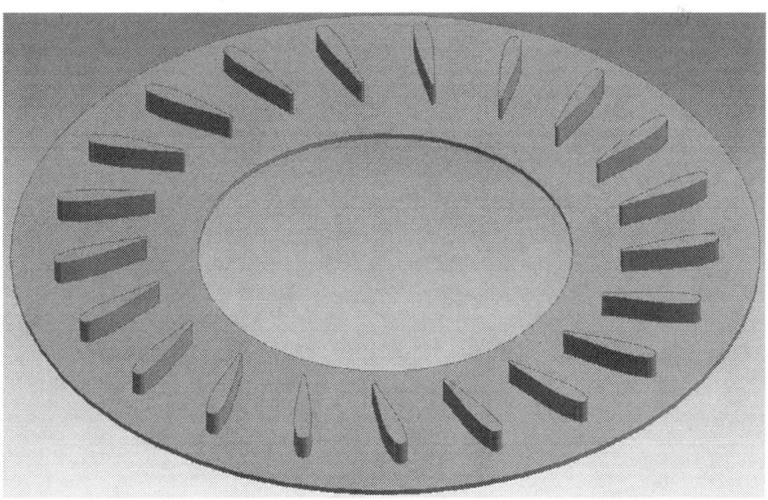

FIGURE 9.35 Stator geometry.

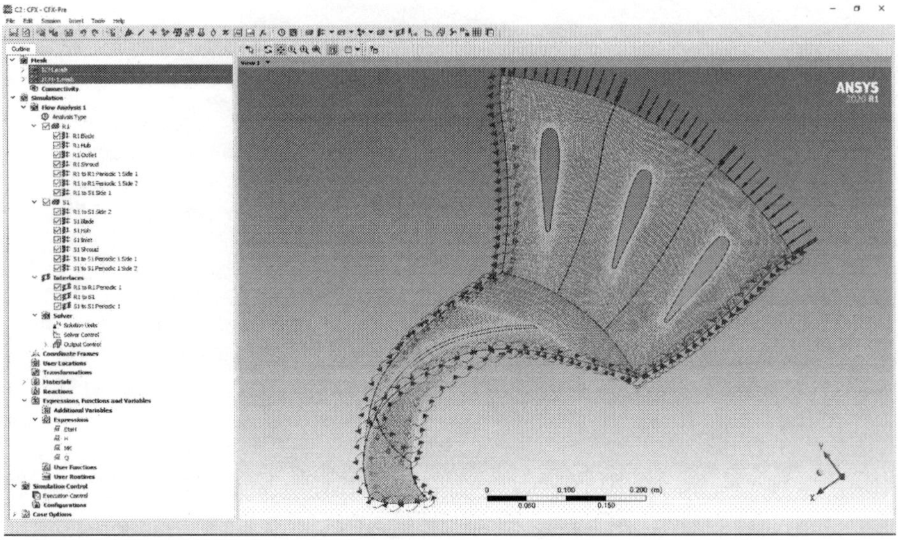

FIGURE 9.36 Computational domain consisting of three stator and one rotor blades in the ANSYS CFX Pre environment.

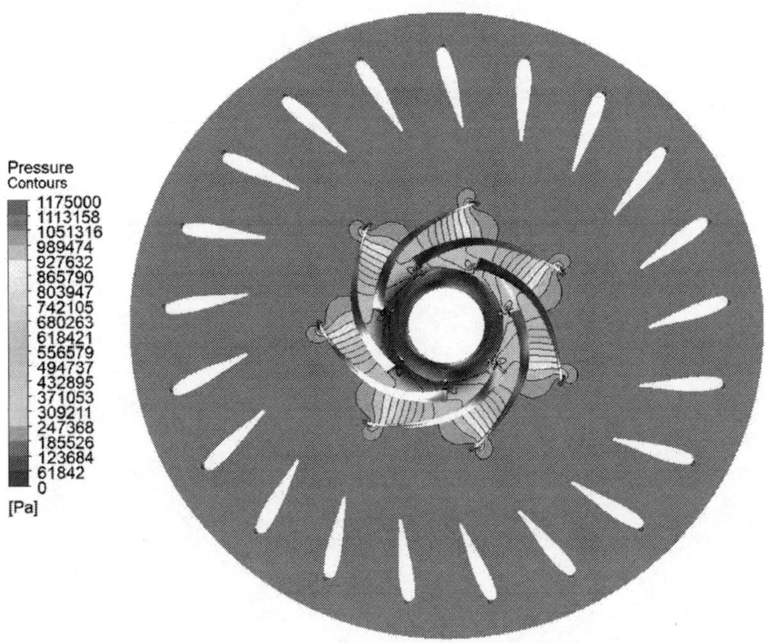

FIGURE 9.37 Pressure distribution in the impeller and the stator.

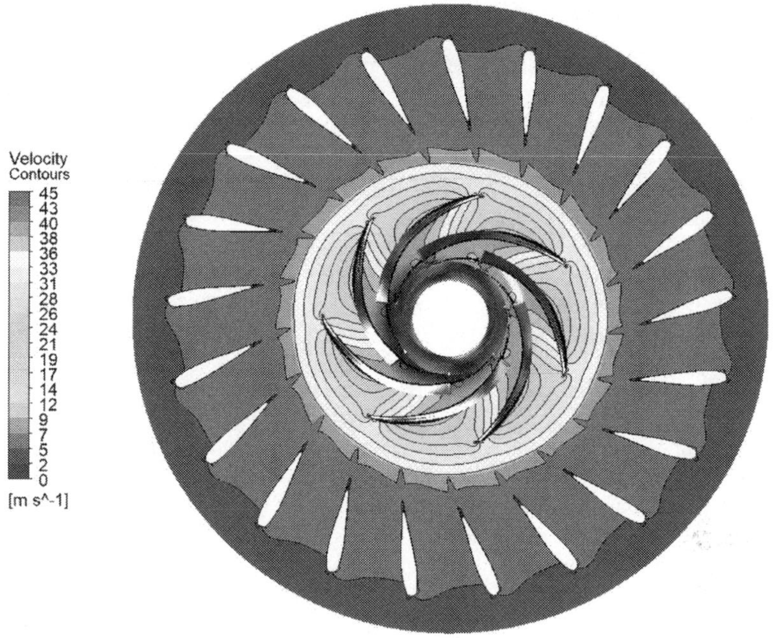

FIGURE 9.38 Velocity distribution inside the impeller and the stator.

one vane of the stator (also with the appropriate overlap in front of and behind the vane) and then create a computational mesh (Figure 9.36). The computational mesh is created in the same way as in the case of the impeller in Example 9.2.

- The final computational domain includes the space around the three vanes of the stator (Figure 9.36). The pitch angle of the volume around the three stator vanes is identical to the pitch angle around one rotor blade. Similarly, in the case of the side walls of the rotor volume, boundary zones of the "periodic" type are defined on the side walls of the stator computational domain.

After all settings have been done and a simulation has been performed, the results can be evaluated. Figures 9.37 and 9.38 show the simulated distribution of static pressure and velocity. As in Example 9.2, the display is performed for all inter-blade spaces of the rotor and stator. The periodicity of the calculated results around the machine axis was applied when creating the display. Figure 9.38 shows the discontinuity in the simulated velocity values in the region of the interface between the rotor and the stator. The main reason is the fact that the equations describing the flow in the stator domain are solved for a stationary coordinate system and in the rotor domain they are solved in a rotating coordinate system. In addition, the interface itself ensures the averaging of the flow field quantities in the circumferential direction (at the location of the interface itself). The calculated head is 50.03 m and the calculated hydraulic efficiency is 81.3%.

9.8 BIBLIOGRAPHY

Alfonsi, G. "On direct numerical simulation of turbulent flows," *Appl. Mech. Rev.* 2011, 64(2), 0802.

Anderson, J. *Computational Fluid Dynamics: The Basics with Applications. 1995,* New York: McGrawHill, 1995.

ANSYS Software Manual CFX Release 12.1, year 2012. Available online. Headquarters: Canonsburg, PA, USA.

Beaudoin, M., Nilsson, H., Page, M., Magnan, R. and Jasak, H. "Evaluation of an improved mixing plane interface for OpenFOAM," Volume 22, Computational and Experimental Techniques, in *IOP Conference Series: Earth and Environmental Science*, 2014. Montreal, Canada.

Davidson, L. *An Introduction to Turbulence Models*. Gothenburg: Department of Thermo and Fluid Dynamics, Chalmers University of Technology, 2003.

Gülich, J. F. *Centrifugal Pumps, fourth edition*. Berlin: Springer, 2000.

Hlbočan, P. "Analysis of the influence of selected geometric parameters on the effective properties of a diagonal pump (in Slovak). Slovak Technical University, 2013.

Izmaylov, R. A., Lopulalan, H. D. and Norimarna, G. S. "Unsteady flow modeling using transient rotor–stator interface," in *Proceedings of the ASME Turbo Expo*, 2013. Volume 6C: Turbomachinery. Turbine Technical Conference and Exposition, San Antonio, Texas, USA.

Liu, M., Liu, M., Dong, L., Ren, Y. and Du, H. "Effects of computational grids and turbulence models on numerical simulation of centrifugal pump with CFD," in *IOP Conference Series: Earth and Environmental Science*, 2012. 26th IAHR Symposium on Hydraulic Machinery and Systems, Volume 15, Session 4: Advances in Computational and Experimental Techniques, Beijing, China.

Menter, F. R., Kuntz, M. and Langtry, R. "Ten years of industrial experience with the SST turbulence model turbulence heat and mass transfer," *Cfd. Spbstu. Ru*, 2003. Published in: Proceedings of the 4th International Symposium on Turbulence, Heat and Mass Transfer, Begell House, Inc., Redding, pp. 625–632.

Menter, F. R., Langtry, R. and Hansen, T. "CFD simulation of turbomachinery flows–Verification, validation and modelling," in P. Neittaanmäki, T. Rossi, K. Majava, O. Pironneau (eds.), *ECCOMAS 2004 – European Congress on Computational Methods in Applied Sciences and Engineering*, 2004. Jyväskylä, Finland.

Perez, J., Chiva, J. S., Segala, W., Morales, R., Negrao, C., Julia, E. and Hernandez, L. "Performance analysis of flow in an impeller-diffuser centrifugal pump using CFD: Simulation and experimental data comparisons," in J. C. F. Pereira and A. Sequeira (eds.), *V European Conference on Computational Fluid Dynamics ECCOMAS CFD 2010*, 2010. Lisbon, Portugal.

Rousseau, P. M., Soulaïmani, A. and Sabourin, M. "Comparison between structured hexahedral and hybrid tetrahedral meshes generated by commercial software for CFD hydraulic turbine analysis," *21st Annual Conference of CFD Society of Canada*, 2013. Sherbrooke, Canada.

Varchola M. and Hlbočan, P. "Hydraulic interaction between an impeller and axial diffuser of a mixed-flow pump (in Slovak)," in *Current Trends in Development of Pumping Machinery,* 2013. Published in: *Applied Mechanics and Materials* 2014; 630, 35–42.

Wilcox, D. C. *Turbulence Modeling for CFD (third edition),"* DCW Industries, 2006.

Witteck, D., Micallef, D. and Mailach, R. "Comparison of transient blade row methods for the CFD analysis of a high-pressure turbine," in *Proceedings of the ASME Turbo Expo*, 2014. Turbine Technical Conference and Exposition. Volume 2D: Turbomachinery. Düsseldorf, Germany.

Index

Note: Figures are indicated by *italics*. Tables are indicated by **bold**.

Made in the USA
Monee, IL
23 May 2026

34934122-2179-4260-8b86-dbe51a2e4791R01